THEORY AND APPLICATIONS OF LINEAR DIFFERENTIAL AND DIFFERENCE EQUATIONS:
A Systems Approach in Engineering

THEORY AND APPLICATIONS OF LINEAR DIFFERENTIAL AND DIFFERENCE EQUATIONS:
A Systems Approach in Engineering

R. M. JOHNSON, B.Sc., F.I.M.A.
Senior Lecturer, Department of Mathematics and Computing
Paisley College of Technology

ELLIS HORWOOD LIMITED
Publishers · Chichester

Halsted Press: a division of
JOHN WILEY & SONS
New York · Chichester · Brisbane · Toronto

First published in 1984 by
ELLIS HORWOOD LIMITED
Market Cross House, Cooper Street, Chichester, West Sussex, PO19 1EB, England

*The publisher's colophon is reproduced from James Gillison's drawing of the
ancient Market Cross, Chichester.*

Distributors:

Australia, New Zealand, South-east Asia:
Jacaranda-Wiley Ltd., Jacaranda Press,
JOHN WILEY & SONS INC.,
G.P.O. Box 859, Brisbane, Queensland 40001, Australia

Canada:
JOHN WILEY & SONS CANADA LIMITED
22 Worcester Road, Rexdale, Ontario, Canada.

Europe, Africa:
JOHN WILEY & SONS LIMITED
Baffins Lane, Chichester, West Sussex, England.

North and South America and the rest of the world:
Halsted Press: a division of
JOHN WILEY & SONS
605 Third Avenue, New York, N.Y. 10016, U.S.A.

© 1984 R.M. Johnson/Ellis Horwood Limited

British Library Cataloguing in Publication Data
Johnson, R. M. (Roy Michael)
Theory and applications of linear differential and difference equations. —
(Ellis Horwood series in mathematics and its applications)
1. Engineering mathematics   2. Differential equations   3. Difference equations
I. Title
620'.001'515354   TA347.D45

Library of Congress Card No. 84–12990

ISBN 0–85312–670–4 (Ellis Horwood Limited — Library Edn.)
ISBN 0–85312–802–2 (Ellis Horwood Limited — Student Edn.)
ISBN 0–470–20106–1 (Halsted Press)

Printed in Great Britain by The Camelot Press, Southampton.

Contents

Contents

Author's preface

This textbook provides a compact treatment of linear differential equations and linear difference equations using transform techniques. It is written as a textbook for engineering, science, computer science, and applied mathematics students in universities and polytechnics. Additionally it will provide a useful starting point for those who require to study more advanced texts, particularly in the field of signal analysis. Students on mathematics courses will find in the book an appreciation of how transform theory is applied in engineering situations.

The level of presentation is appropriate to second year degree mathematics students or to readers who are not specialists in mathematics, such as engineers and scientists, who study continuous and discrete linear systems. Such readers will require a mathematical background approximately equivalent to first year university mathematics.

The transform techniques used in the book are developed to encourage readers to think in terms of transfer functions and block diagrams rather than equations. An important relationship between the transform variables and frequency is established, and system stability is considered. Examples are chosen from the fields of Electrical Engineering, Mechanical Engineering, Civil Engineering, and Control Engineering. The book is probably unique in the way it uses frequency domain analysis to underline the similarities between differential equations and difference equations.

The development of microelectronic technology has given new emphasis to the analysis and filtering of discrete signals. The chapter on digital filters will be particularly useful to those whose background has been mainly in continuous systems.

PART I (Chapters 1–4). After a brief introduction to Fourier series, the Laplace transform is defined and applied to the solution of linear differential equations. Block diagram notation and transfer functions are introduced, and the language of control system analysis is used. Special emphasis is placed on the frequency response function and its application in the study of steady-state oscillations. An introduction is given to the concept of analog filtering and to the use of Bode diagrams. Chapter 4 considers differential equations with piecewise continuous forcing functions and serves as an introduction to sampling devices.

PART II (Chapters 5–8). Consideration of sampling devices leads to the definition of the z-transform which is then applied to the solution of linear difference equations. Again transfer functions are defined and block diagram notation is used. The use of transfer functions allows the similarities between differential equations and difference equations to be highlighted. In particular the methods for obtaining a system's steady-state output from its transfer function are compared. z-transform techniques are applied to simple sampled-data systems and to digital systems with reconstructed outputs. In the final chapter digital filters are introduced and simple design algorithms are established so that the performance of a given analog filter may be copied.

The "dot" notation, $\dot{x} \equiv dx/dt$, is used throughout the book; other notations are defined as they occur in the text. All system inputs are taken to be zero for $t < 0$.

The author is indebted to the many students at Paisley College who, in recent years, have been on the receiving end of much of the material contained in this textbook.

Particular thanks are expressed to Madeleine Stafford for the considerable task of typing and correcting the manuscript.

Finally I am grateful to the Series Editor, Professor G. Bell, to the publishers' referee, and to the staff of Ellis Horwood Limited for their valuable assistance and encouragement.

R. M. Johnson

Paisley 1984

Part 1
Continuous systems

CHAPTER 1

An approach to the Laplace transform

1.1 INTRODUCTION

The Laplace transform is a powerful mathematical tool for problems arising from the study of continuous systems. The term 'continuous systems' is taken to imply systems which can be modelled by ordinary differential equations, for example

(i)   a control system which positions a missile fin to achieve a certain lateral acceleration,
(ii) a crane where the position of the load is controlled by the application of hydraulic motors,
(iii) a structure subject to vibration.

The variables in these examples, position, acceleration, pressure, force, displacement, are *continuous* variables which can take any value within some specified range.

In Part 1 of this book we will apply Laplace transforms to *linear* continuous systems, that is systems described by linear differential equations. This can be done by accepting the mathematical definition of a Laplace transform as a starting point and turning directly to Chapter 2. This first chapter approaches the idea of a Laplace transform by considering the frequency characteristics of a function of time, and attempts to show the important relationship between the Laplace variable and frequency.

1.2 THE FOURIER SERIES OF A PERIODIC FUNCTION

A periodic function $f(t)$ satisfying certain conditions may be expressed as an infinite series which is a linear combination of sine and cosine functions whose frequencies are multiples of the fundamental frequency $\omega_0 = \dfrac{2\pi}{L}$, where L is the period of $f(t)$. The infinite series is known as the **Fourier series expansion** of $f(t)$ and takes the form

$$f(t) = \frac{1}{2}a_0 + \sum_{n=1}^{\infty} (a_n \cos n\omega_0 t + b_n \sin n\omega_0 t). \tag{1.1}$$

where the constants a_n and b_n are given by

$$a_n = \frac{\omega_0}{\pi} \int_{-\pi/\omega_0}^{\pi/\omega_0} f(t) \cos n\omega_0 t \, dt. \tag{1.2}$$

$$b_n = \frac{\omega_0}{\pi} \int_{-\pi/\omega_0}^{\pi/\omega_0} f(t) \sin n\omega_0 t \, dt. \tag{1.3}$$

$$\left(\text{note that } \frac{\pi}{\omega_0} = \frac{L}{2} \right)$$

Assuming that the Fourier series expansion exists for a given function $f(t)$, then equations (1.2) and (1.3) follow immediately from the orthogonality of the functions $\{\sin n\omega_0 t, \cos n\omega_0 t\}$ over the interval $\left[-\dfrac{\pi}{\omega_0}, \dfrac{\pi}{\omega_0} \right]$. That is,

when $n \neq m$, m and n integers

$$\int_{-\pi/\omega_0}^{\pi/\omega_0} \cos n\omega_0 t \cos m\omega_0 t \, dt = 0$$

$$\int_{-\pi/\omega_0}^{\pi/\omega_0} \sin n\omega_0 t \sin m\omega_0 t \, dt = 0,$$

and for all integers m, n

$$\int_{-\pi/\omega_0}^{\pi/\omega_0} \sin n\omega_0 t \cos n\omega_0 t \, dt = 0.$$

Sufficient conditions for the existence of the series (1.1) are that the function $f(t)$ is bounded and has a finite number of discontinuities and a finite number of maxima and minima in the interval $\left[-\dfrac{\pi}{\omega_0}, \dfrac{\pi}{\omega_0} \right]$. If $t = t_1$ is a point where $f(t)$ is continuous then the series converges to $f(t_1)$; if $t = t_2$ is a point where $f(t)$ is discontinuous then the series converges to $\frac{1}{2}\{f(t_2{}^+) + f(t_2{}^-)\}$, where $t_2 - \varepsilon < t_2{}^- < t_2 < t_2{}^+ < t_2 + \varepsilon$ for arbitrarily small ε. For a more detailed treatment of Fourier series see Kreyszig (1979).

Example 1.1

Obtain the Fourier series expansion of the function

$$f(t) = \begin{cases} 0, & -\pi \leqslant t < 0 \\ 1, & 0 \leqslant t < \pi \end{cases}, \quad f(t + 2\pi) = f(t).$$

The function is shown in Fig. 1.1.

The period of the function is $L = 2\pi$, therefore $\omega_0 = 1$.

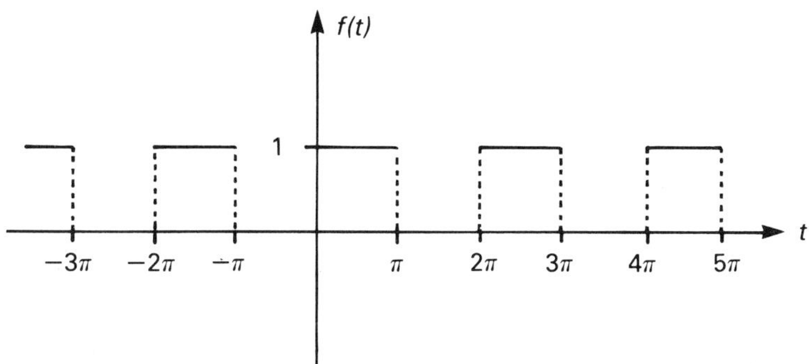

Fig. 1.1

Equations (1.2) and (1.3) give

$$a_0 = \frac{1}{\pi} \int_{-\pi}^{\pi} f(t)\, dt$$

$$= \frac{1}{\pi} \int_{0}^{\pi} dt$$

$$= 1$$

$$a_n = \frac{1}{\pi} \int_{-\pi}^{\pi} f(t) \cos nt\, dt \quad \text{for} \quad n = 1, 2, 3, \ldots,$$

$$= \frac{1}{\pi} \int_{0}^{\pi} \cos nt\, dt$$

$$= 0$$

$$b_n = \frac{1}{\pi} \int_{-\pi}^{\pi} f(t) \sin nt\, dt \quad \text{for} \quad n = 1, 2, 3, \ldots,$$

$$= \frac{1}{\pi} \int_{0}^{\pi} \sin nt\, dt$$

$$= \begin{cases} 2/\pi n, & \text{when } n \text{ is odd} \\ 0, & \text{when } n \text{ is even.} \end{cases}$$

Substituting these results into equation (1.1) gives the required series,

$$f(t) = 0.5 + \frac{2}{\pi} \left\{ \sin t + \frac{\sin 3t}{3} + \frac{\sin 5t}{5} + \cdots \right\}.$$

Note that the above series converges to 1 for $0 < t < \pi$, converges to 0 for

$-\pi < t < 0$, and converges (clearly) to 0.5 when $t = 0$. For example, putting $t = \dfrac{\pi}{2}$,

$$1 = 0.5 + \frac{2}{\pi}\left\{ 1 - \frac{1}{3} + \frac{1}{5} - \frac{1}{7} + \cdots \right\},$$

that is, $\displaystyle\sum_{r=0}^{\infty} \frac{(-1)^r}{(2r+1)} = \frac{\pi}{4}.$

Example 1.2

The function $f(t) = t$ is defined in the interval $0 \leqslant t \leqslant l$. Obtain an infinite series expansion of $f(t)$ of the form

$$f(t) = \frac{1}{2}a_0 + \sum_{n=1}^{\infty} a_n \cos \frac{n\pi}{l} t, \quad 0 \leqslant t \leqslant l.$$

Hence show that

$$\sum_{r=0}^{\infty} \frac{1}{(2r+1)^2} = \frac{\pi^2}{8}.$$

We consider the *even* function $f_{\mathrm{E}}(t)$ defined as follows

$$f_{\mathrm{E}}(t) = \begin{cases} t, & 0 \leqslant t < l \\ -t, & -l \leqslant t < 0 \end{cases}, \quad f_{\mathrm{E}}(t + 2l) = f_{\mathrm{E}}(t).$$

The function $f_{\mathrm{E}}(t)$, shown in Fig. 1.2(a), is a periodic function of period $2l$ and coincides with the given function $f(t)$ in the interval $0 \leqslant t \leqslant l$. Since it is an even function, $f_{\mathrm{E}}(t)$ will have a Fourier series expansion which contains only cosine terms. That is,

$$f_{\mathrm{E}}(t) = \frac{1}{2}a_0 + \sum_{n=1}^{\infty} a_n \cos n\omega_0 t, \quad \omega_0 = \frac{\pi}{l}.$$

Therefore the function $f(t)$ will be represented by this series in the interval $0 \leqslant t \leqslant l$. ($f_{\mathrm{E}}(t)$ is continuous for all t.)

Equation (1.2) gives

$$a_0 = \frac{1}{l} \int_{-l}^{l} f_{\mathrm{E}}(t)\, dt$$

$$= \frac{2}{l} \int_{0}^{l} t\, dt, \text{ since } f_{\mathrm{E}}(t) \text{ is even,}$$

$$= l$$

$$a_n = \frac{1}{l} \int_{-l}^{l} f(t) \cos \frac{n\pi}{l} t \, dt, \quad n = 1, 2, 3, \ldots$$

$$= \frac{2}{l} \int_{0}^{l} t \cos \frac{n\pi}{l} t \, dt$$

$$= \frac{2}{l} \left\{ \frac{l^2}{n\pi} \sin n\pi + \frac{l^2}{n^2 \pi^2} (\cos n\pi - 1) \right\}.$$

Therefore $a_n = \begin{cases} 0, \, n = 2, 4, 6, \ldots \\ \dfrac{-4l}{n^2 \pi^2}, \, n = 1, 3, 5, \ldots \end{cases}$

and $f(t) = \dfrac{l}{2} - \dfrac{4l}{\pi^2} \left\{ \cos \dfrac{\pi}{l} t + \dfrac{1}{3^2} \cos \dfrac{3\pi}{l} t + \dfrac{1}{5^2} \cos \dfrac{5\pi}{l} t + \ldots \right\}.$

Setting $t = 0$ gives

$$0 = \frac{l}{2} - \frac{4l}{\pi^2} \left\{ 1 + \frac{1}{3^2} + \frac{1}{5^2} + \ldots \right\},$$

that is, $\displaystyle\sum_{r=0}^{\infty} \frac{1}{(2r+1)^2} = \frac{\pi^2}{8}.$

Note that if we require a *sine* series for $f(t) = t, 0 \leqslant t \leqslant l$, then it is necessary to consider the periodic *odd* function,

$$f_o(t) = t, \, -l \leqslant t < l, \, f_o(t + 2l) = f_o(t).$$

The Fourier series expansion of $f_o(t)$ contains only sine terms, that is, $a_n = 0$ for $n = 0, 1, 2, 3, \ldots$ (See Fig. 1.2(b) and Problem 3.)

Problems

(1) Obtain the Fourier series expansion of the function

$$f(t) = \begin{cases} \pi + t, \, -\pi \leqslant t < 0 \\ \pi - t, \quad 0 \leqslant t < \pi \end{cases}, \, f(t + 2\pi) = f(t).$$

(2) The output of a half-wave rectifier is given by

$$f(t) = \begin{cases} E \sin \dfrac{2\pi}{T} t, \, 0 \leqslant t < \dfrac{T}{2} \\ 0, \quad \dfrac{T}{2} < t < T \end{cases}, f(t + T) = f(t).$$

Show that $f(t) = \dfrac{E}{\pi} + \dfrac{E}{2} \sin \dfrac{2\pi}{T} t - \dfrac{2E}{\pi} \left\{ \dfrac{\cos \dfrac{4\pi}{T} t}{1.3} + \dfrac{\cos \dfrac{8\pi}{T} t}{3.5} + \ldots \right\}.$

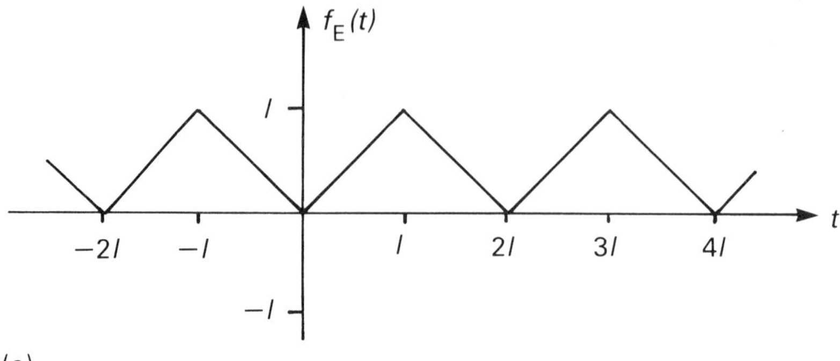

(a)

(b)

Fig. 1.2 (a) Even extension of $f(t)$, (b) Odd extension of $f(t)$.

(3) The function $f(t) = t$ is defined on the interval $0 \leqslant t \leqslant l$. Obtain an infinite series expansion of $f(t)$ in the form

$$f(t) = \sum_{n=1}^{\infty} b_n \sin \frac{n\pi}{l} t.$$

(4) By consideration of the Fourier series for $f(t) = t^2$, $-\pi \leqslant t \leqslant \pi$, show that

(i) $\quad \dfrac{\pi^2}{6} = 1 + \dfrac{1}{4} + \dfrac{1}{9} + \dfrac{1}{16} + \cdots$

(ii) $\quad \dfrac{\pi^2}{12} = 1 - \dfrac{1}{4} + \dfrac{1}{9} - \dfrac{1}{16} + \cdots$

(5) Obtain the Fourier series for $f(t) = e^{-a|t|}$, $a > 0$, $-L \leqslant t \leqslant L$ and deduce that

$$\sum_{r=1}^{\infty} \left\{ a^2 + (2r-1)^2 \frac{\pi^2}{L^2} \right\}^{-1} = \frac{L}{4a} \left\{ \frac{1 - e^{-aL}}{1 + e^{-aL}} \right\}.$$

Answers to problems

(1) $\dfrac{\pi}{2} + \dfrac{4}{\pi}\left\{\cos t + \dfrac{1}{9}\cos 3t + \dfrac{1}{25}\cos 5t + \ldots\right\}$

(3) $\dfrac{2l}{\pi}\left\{\sin\dfrac{\pi}{l}t - \dfrac{1}{2}\sin\dfrac{2\pi}{l}t + \dfrac{1}{3}\sin\dfrac{3\pi}{l}t - \ldots\right\}, \quad 0 \leqslant t < l.$

(5) $\dfrac{1}{aL}(1 - e^{-aL}) + \dfrac{2a}{L}\sum_{n=1}^{\infty}(1 - (-1)^n e^{-aL})\left(a^2 + \dfrac{n^2\pi^2}{L^2}\right)^{-1}\cos\dfrac{n\pi}{L}t.$

1.3 COMPLEX FORM OF FOURIER SERIES

The sequence of numbers $\{a_0, a_1, a_2, \ldots, b_1, b_2, \ldots\}$ is an alternative way of defining the function $f(t)$ since given these numbers we can reconstruct the function using the Fourier series, equation (1.1). Writing the series as

$$f(t) = \frac{1}{2}a_0 + \sum_{n=1}^{\infty} r_n \cos(n\omega_0 t + \phi_n)$$

where $r_n = \sqrt{(a_n^2 + b_n^2)}$, $n = 1, 2, 3, \ldots$, each number r_n provides a measure of the contribution to the periodic function $f(t)$ from the frequency $n\omega_0$. This information can be presented graphically in the form of a **Discrete amplitude spectrum**, see Fig. 1.3. In particular, the discrete amplitude spectrum for Example 1.1 is shown in Fig. 1.4.

An alternative, and more usual, representation of the discrete amplitude spectrum is related to the *complex form* of the Fourier series expansion.

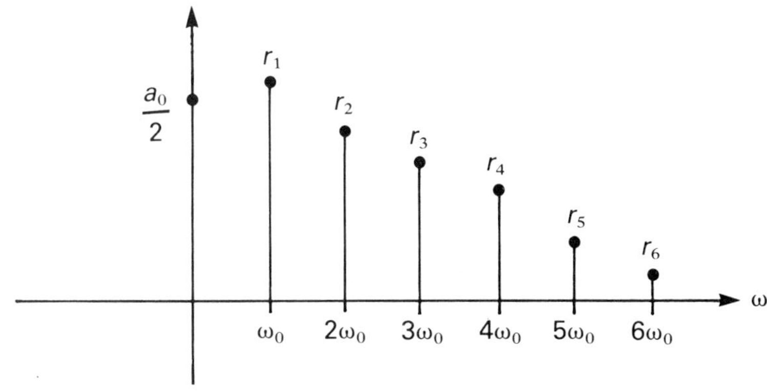

Fig. 1.3

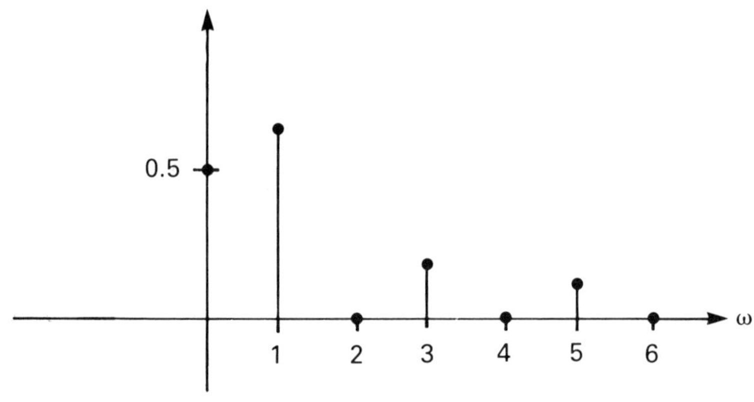

Fig. 1.4

Making the substitutions

$$\cos n\omega_0 t = \frac{1}{2}\left(\exp\left(jn\omega_0 t\right) + \exp\left(-jn\omega_0 t\right)\right)\dagger$$

$$\sin n\omega_0 t = \frac{1}{2j}\left(\exp\left(jn\omega_0 t\right) - \exp\left(-jn\omega_0 t\right)\right)$$

where $j^2 = -1$ and $\exp(u)$ denotes e^u, equation (1.1) becomes

$$f(t) = \frac{1}{2}a_0 + \frac{1}{2}\sum_{n=1}^{\infty}\left\{(a_n - jb_n)\exp\left(jn\omega_0 t\right) + (a_n + jb_n)\exp\left(-jn\omega_0 t\right)\right\}. \tag{1.4}$$

If we define

$$c_0 = \frac{1}{2}a_0$$

$$c_n = \frac{1}{2}(a_n - jb_n), \quad n = 1, 2, 3, \ldots$$

$$c_{-n} = \frac{1}{2}(a_n + jb_n), \quad n = 1, 2, 3, \ldots,$$

then equation (1.4) can be written in the *complex form*

$$f(t) = \sum_{n=-\infty}^{n=\infty} c_n \exp\left(jn\omega_0 t\right). \tag{1.5}$$

where

$$c_n = \frac{\omega_0}{2\pi}\int_{-\pi/\omega_0}^{\pi/\omega_0} f(t)\exp\left(-jn\omega_0 t\right)dt. \tag{1.6}$$

$\dagger$ For convenience in printing the notation $\exp(\ )$ is frequently used for $e^{(\ )}$.

Equation (1.6) follows immediately from equation (1.2), equation (1.3), and the definition of c_n.

The sequence of numbers $\{c_n, n = 0, \pm 1, \pm 2, \pm 3, \ldots\}$ is an alternative method of defining the function $f(t)$, since the function can be reconstructed from these coefficients using the complex form Fourier series, equation (1.5).

Note that, if $f(t)$ is a real function, then

$$c_{-n} = \bar{c}_n \qquad \text{and}$$

$$|c_n| = |c_{-n}| = \frac{1}{2}\sqrt{a_n^2 + b_n^2}$$

$$= \frac{1}{2}r_n,$$

that is, $|c_n|$ is one half of the contribution to $f(t)$ from the frequency $n\omega_0$. This information can be represented graphically as a discrete amplitude spectrum, see Fig. 1.5. Note that the discrete amplitude spectrum is symmetrical about $\omega = 0$ for a real function $f(t)$. The amplitude contribution to $f(t)$ from the frequency $n\omega_0$ is obtained by adding the height of the spectrum at $n\omega_0$ and at the 'negative frequency' $-n\omega_0$, that is,

$$r_n = |c_n| + |c_{-n}|$$

$$= 2|c_n| \text{ for a real function.}$$

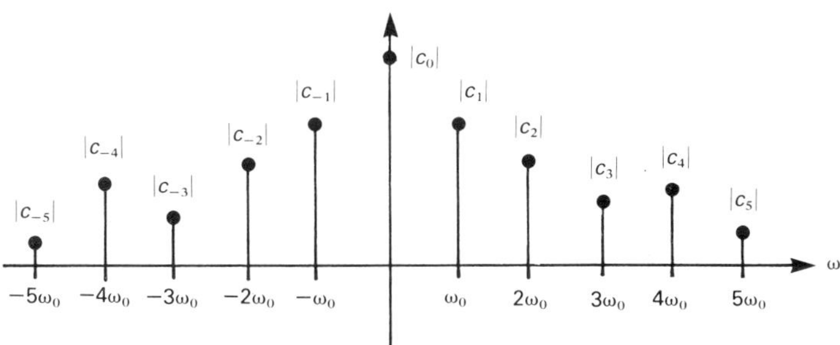

Fig. 1.5 Discrete amplitude spectrum for real periodic function.

Example 1.3

Obtain the complex form Fourier series for the function

$$f(t) = \begin{cases} 1, & 0 \leqslant t < \pi \\ 0, & \pi \leqslant t < 2\pi \end{cases}, \quad f(t + 2\pi) = f(t)$$

and sketch its discrete amplitude spectrum. (This function was considered in Example 1.1.)

From equation (1.6)
$$c_n = \frac{1}{2\pi} \int_{-\pi}^{\pi} f(t)\,e^{-jnt}\,dt$$

$$= \frac{1}{2\pi} \int_{0}^{\pi} e^{-jnt}\,dt$$

$$= -\frac{1}{2\pi jn} \left[e^{-jnt} \right]_{0}^{\pi}, \quad n \neq 0$$

$$= \frac{j}{2\pi n} (e^{-jn\pi} - 1), \quad n \neq 0$$

$$c_0 = \frac{1}{2\pi} \int_{0}^{\pi} dt = \frac{1}{2}.$$

Therefore
$$c_n = \begin{cases} \dfrac{1}{2}, & n = 0 \\[2ex] \dfrac{-j}{\pi n}, & n = \pm 1, \pm 3, \pm 5, \ldots \\[2ex] 0, & n = \pm 2, \pm 4, \pm 6, \ldots \end{cases}$$

The complex form Fourier series can be written as

$$f(t) = \frac{1}{2} - \frac{j}{\pi} \sum_{r=-\infty}^{r=\infty} \frac{e^{j(2r-1)t}}{(2r-1)}.$$

The discrete amplitude spectrum is shown in Fig. 1.6.

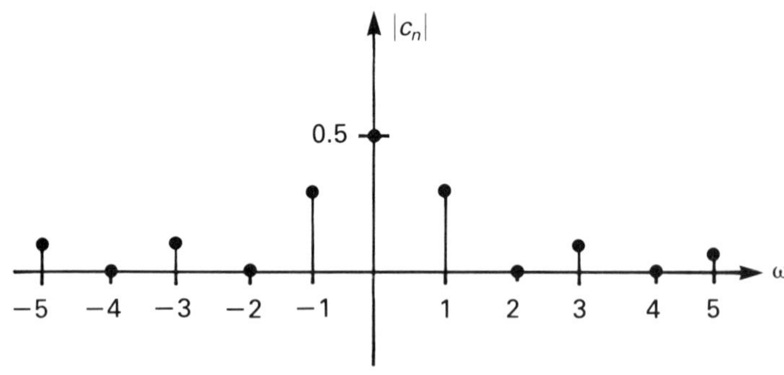

Fig. 1.6 Discrete amplitude spectrum for Example 1.3.

Problems

(1) Show that the complex form of the Fourier series of the function $f(t) = e^t$ when $-\pi < t < \pi$, $f(t + 2\pi) = f(t)$ is given by

$$f(t) = \frac{2\sinh\pi}{\pi} \sum_{n=-\infty}^{\infty} (-1)^n \frac{1+jn}{1+n^2} e^{jnt}.$$

(2) From the series in Problem 1, deduce that

$$f(t) = \frac{2\sinh\pi}{\pi} \left\{ \frac{1}{2} - \frac{1}{2}(\cos t - \sin t) + \frac{1}{5}(\cos 2t - 2\sin 2t) - \ldots \right\}.$$

(3) Given that $f(t) = \sum_{n=-\infty}^{\infty} c_n e^{jn\omega t}$, show that when $f(t)$ is an odd function c_n is purely imaginary, and when $f(t)$ is an even function c_n is real.

1.4 EXTENSION TO NON-PERIODIC FUNCTIONS—THE FOURIER TRANSFORM

We consider a non-periodic function as the limit of a periodic function when the period is allowed to increase without bound. For example, consider the function

$$f(t) = \begin{cases} 1 \text{ when } 0 \leqslant t < 1 \\ 0 \text{ otherwise} \end{cases},$$

which is shown in Fig. 1.7(a). This function can be considered as $\lim_{T \to \infty} f_p(t)$

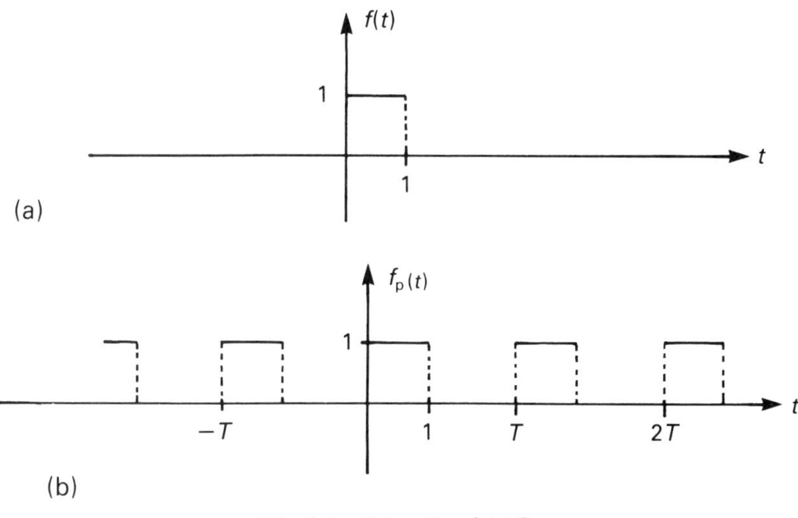

(a)

(b)

Fig. 1.7 $f(t) = \lim_{T \to \infty} \{f_p(t)\}.$

where $f_p(t)$ is the periodic function shown in Fig. 1.7(b) and defined by

$$f_p(t) = \begin{cases} 1 & \text{when } 0 \leqslant t < 1 \\ 0 & \text{when } 1 \leqslant t < T \end{cases}, \quad f_p(t+T) = f_p(t).$$

In general we can express a periodic function $f_p(t)$, with period T, as a complex form Fourier series according to equations (1.5) and (1.6). Substituting for c_n in equation (1.5) gives

$$f_p(t) = \sum_{n=-\infty}^{\infty} \left\{ \frac{\omega_0}{2\pi} \int_{-\pi/\omega_0}^{\pi/\omega_0} f(u)e^{-jn\omega_0 u}\, du \right\} e^{jn\omega_0 t}. \tag{1.7}$$

where $\omega_0 = \dfrac{2\pi}{T}$ and the dummy variable in equation (1.6) has been changed from t to u to avoid confusion.

Defining $\omega = n\omega_0$ to be a *continuous* frequency variable and $\Delta\omega = \omega_0$, the difference between adjacent ordinates on the discrete amplitude spectrum of $f_p(t)$, equation (1.7) can be written

$$f_p(t) = \frac{1}{2\pi} \sum_{n=-\infty}^{\infty} \left\{ \int_{-\pi/\Delta\omega}^{\pi/\Delta\omega} f(u)e^{-j\omega u}\, du \right\} e^{j\omega t}\, \Delta\omega.$$

Now letting the period $T \to \infty$, that is, $\omega_0 = \Delta\omega \to 0$ and using a fundamental theorem of integral calculus (a definite integral is the limit of a summation) we obtain

$$f(t) = \frac{1}{2\pi} \int_{-\infty}^{\infty} \left\{ \int_{-\infty}^{\infty} f(u)e^{-j\omega u}\, du \right\} e^{j\omega t}\, d\omega. \tag{1.8}$$

This can be written

$$f(t) = \frac{1}{2\pi} \int_{-\infty}^{\infty} F(j\omega)e^{j\omega t}\, d\omega. \tag{1.9}$$

where

$$F(j\omega) = \int_{-\infty}^{\infty} f(t)e^{-j\omega t}\, dt. \tag{1.10}$$

(changing the dummy variable back to t).

Equation (1.10) defines the function $F(j\omega)$ which is called the **Fourier transform** of $f(t)$. Equation (1.9) is the **inversion formula**, that is, $f(t)$ is the **inverse Fourier transform** of $F(j\omega)$. The equations (1.9) and (1.10) are called a **Fourier transform pair** and together are equivalent to equation (1.8) which is known as the **Fourier integral formula**. Other definitions for Fourier transform and inversion formula are given in the literature, but taken *together* will reduce to equation (1.8). (An alternative is to change the frequency variable to $v = \dfrac{\omega}{2\pi}$ (Hertz) which makes equations (1.9) and (1.10) the same apart from the sign of j). For a more detailed treatment of the Fourier transform see Oppenheim & Willsky (1983).

The Fourier integral formula is the equivalent for non-periodic functions of the Fourier series expansion, and the function $F(j\omega)$ corresponds to the set of Fourier coefficients $\{c_n; n = 0, \pm 1, \pm 2, \ldots\}$. $F(j\omega)$ is a function of ω and is referred to as the **spectrum** of $f(t)$. The **amplitude spectrum** $|F(j\omega)|$ is now a continuous function of ω, see Fig. 1.8. By writing

$$e^{-j\omega t} = \cos \omega t - j \sin \omega t$$

in equation (1.10), it is easy to show that $|F(j\omega)| = |F(-j\omega)|$, that is, the amplitude spectrum is an *even* function of ω.

The function $F(j\omega)$ is an alternative way of representing the function $f(t)$ since, given $F(j\omega)$ we can reconstruct the function $f(t)$ by using the inversion formula. The spectrum $F(j\omega)$ is called the representation of $f(t)$ in the **frequency domain**. Assuming that the integral in equation (1.10) converges, then $F(j\omega)$ is unique for a given $f(t)$. However, not all functions $f(t)$ can be expressed in the form of the Fourier integral formula, that is, the Fourier transform $F(j\omega)$ does not exist for certain functions $f(t)$. *Sufficient* conditions for the existence of $F(j\omega)$ are:

(i) $f(t)$ has a finite number of maxima and minima and discontinuities in any finite interval,

(ii) $\displaystyle\int_{-\infty}^{\infty} |f(t)|\, dt$ is finite.

We define the **energy** of a function $f(t)$ as $\int_{-\infty}^{\infty} \{f(t)\}^2\, dt$, and the conditions (i) and (ii) include all useful functions with *finite* energy.

From equation (1.9) it is easy to show that the energy of $f(t)$ can also be expressed as

$$\frac{1}{\pi} \int_{0}^{\infty} |F(j\omega)|^2\, d\omega.$$

It follows that the energy of $f(t)$ contained in the frequency band $\omega_1 < \omega < \omega_2$ is

$$\frac{1}{\pi} \int_{\omega_1}^{\omega_2} |F(j\omega)|^2\, d\omega.$$

Many simple functions, however, do not possess finite energy, and condition (ii) is not satisfied. For example, $f(t) = 1, f(t) = \sin t$. We define the **power** of a function as energy per unit time, that is,

$$\lim_{T \to \infty} \frac{1}{T} \int_{-T/2}^{T/2} \{f(t)\}^2\, dt.$$

Functions with *infinite* energy and *finite* power do not possess Fourier

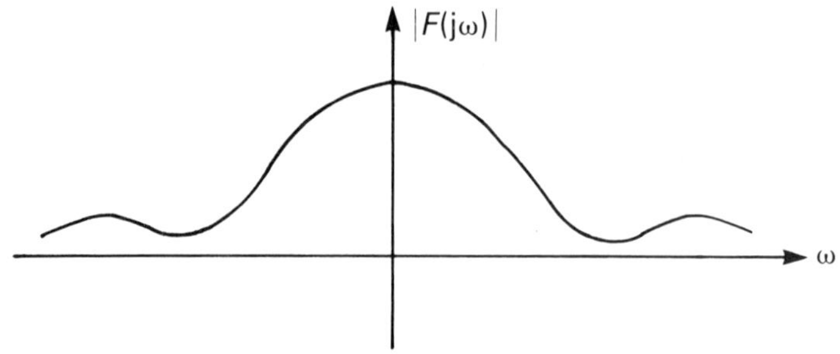

Fig. 1.8 Amplitude spectrum for real non-periodic function.

transforms in terms of elementary functions, but in certain cases the transforms can be expressed in terms of impulse functions, see Example 1.5. For further information on power signals and energy signals see Gabel & Roberts (1973).

Example 1.4

Find the Fourier transform of the function shown in Fig. 1.7(a). That is,

$$f(t) = \begin{cases} 1 & \text{when } 0 \leqslant t < 1 \\ 0 & \text{otherwise.} \end{cases}$$

Hence show that

$$\int_0^\infty \frac{\sin \omega}{\omega}\, d\omega = \frac{\pi}{2}$$

$$F(j\omega) = \int_{-\infty}^\infty f(t)\,e^{-j\omega t}\, dt = \int_0^1 e^{-j\omega t}\, dt$$

$$= -\frac{1}{j\omega}\left[e^{-j\omega t}\right]_0^1$$

$$= \frac{1 - e^{-j\omega}}{j\omega}.$$

From the inversion formula, equation (1.9),

$$f(t) = \frac{1}{2\pi}\int_{-\infty}^\infty \left(\frac{1 - e^{-j\omega}}{j\omega}\right) e^{j\omega t}\, d\omega.$$

This integral representation of $f(t)$ will converge to the value of $f(t)$ at points where $f(t)$ is continuous, but at a point of discontinuity of $f(t)$ the integral will converge to $\frac{1}{2}\{f(t^+)+f(t^-)\}$. Therefore when $t = 0$, we have

$$\frac{1}{2}\left\{1+0\right\} = \frac{1}{2\pi}\int_{-\infty}^{\infty}\left(\frac{1-e^{-j\omega}}{j\omega}\right)d\omega$$

$$= \frac{1}{2\pi}\int_{-\infty}^{\infty}\left(\frac{1-\cos\omega}{j\omega}\right)d\omega + \frac{1}{2\pi}\int_{-\infty}^{\infty}\frac{\sin\omega}{\omega}\,d\omega$$

$$= \frac{1}{\pi}\int_{0}^{\infty}\frac{\sin\omega}{\omega}\,d\omega,$$

since the first integrand is an odd function of ω, and the second integrand is an even function of ω.

Therefore $\displaystyle\int_{0}^{\infty}\frac{\sin\omega}{\omega}\,d\omega = \frac{\pi}{2}.$

(Note how the Fourier integral formula has been used to evaluate an improper integral in a similar way to the use of Fourier series to sum the infinite series in Examples 1.1 and 1.2.)

Example 1.5

We define the following functions

(i) The *symmetric pulse function* $P_\alpha(t)$,

$$P_\alpha(t) = \begin{cases} 1 \text{ when } -\frac{\alpha}{2} \leqslant t < \frac{\alpha}{2} \\ 0 \text{ otherwise} \end{cases}.$$

(ii) The *delayed pulse function* $P_\alpha(t-T)$,

$$P_\alpha(t-T) = \begin{cases} 1 \text{ when } T-\frac{\alpha}{2} \leqslant t < T+\frac{\alpha}{2} \\ 0 \text{ otherwise} \end{cases}.$$

(iii) The *unit impulse function* $\delta(t)$,

$$\delta(t) = \lim_{\alpha \to 0}\frac{1}{\alpha}P_\alpha(t).$$

(iv) The *delayed impulse function* $\delta(t-T)$,

$$\delta(t-T) = \lim_{\alpha \to 0}\frac{1}{\alpha}P_\alpha(t-T).$$

Obtain the Fourier transform of $P_\alpha(t-T)$ and $\delta(t)$. Deduce that the Fourier transform of the function $f(t) = 1$ is $2\pi\delta(\omega)$.

By considering the Fourier transform of $\delta(t-T)+\delta(t+T)$ obtain the Fourier transform of the function $\cos\omega_0 t$.

———————————

For the function $P_\alpha(t-T)$, the Fourier transform is given by

$$\int_{T-\frac{\alpha}{2}}^{T+\frac{\alpha}{2}} e^{-j\omega t}\,dt = \frac{\exp(-j\omega(T-\frac{\alpha}{2})) - \exp(-j\omega(T+\frac{\alpha}{2}))}{j\omega}$$

$$= \frac{2}{\omega}e^{-j\omega T}\sin\frac{\omega\alpha}{2}.$$

Setting $T=0$, the Fourier transform of $P_\alpha(t)$ is simply $\dfrac{2}{\omega}\sin\dfrac{\omega\alpha}{2}$, and the Fourier transform of $\delta(t)$ is given by

$$\lim_{\alpha\to 0}\left\{\frac{2}{\omega\alpha}\sin\frac{\omega\alpha}{2}\right\} = \lim_{\theta\to 0}\left\{\frac{\sin\theta}{\theta}\right\}, \quad \text{where } \theta = \frac{\omega\alpha}{2}$$

$$= 1.$$

Since the Fourier transform of $\delta(t)$ is 1, the inversion formula gives

$$\delta(t) = \frac{1}{2\pi}\int_{-\infty}^{\infty} e^{j\omega t}\,d\omega.$$

By simply interchanging t and the dummy variable ω, we obtain after a sign change substitution

$$2\pi\delta(\omega) = \int_{-\infty}^{\infty} e^{-j\omega t}\,dt.$$

Therefore the Fourier transform of the function $f(t)=1$ is given by $2\pi\delta(\omega)$.

The Fourier transform of $\delta(t-T)+\delta(t+T)$ is given by

$$\lim_{\alpha\to 0}\left\{\frac{2}{\alpha\omega}e^{-j\omega T}\sin\frac{\omega\alpha}{2} + \frac{2}{\alpha\omega}e^{j\omega T}\sin\frac{\omega\alpha}{2}\right\} = e^{j\omega T}+e^{-j\omega T}$$

$$= 2\cos\omega T$$

The inversion formula gives

$$\delta(t-T)+\delta(t+T) = \frac{1}{2\pi}\int_{-\infty}^{\infty} 2\cos\omega T\cdot e^{j\omega t}\,d\omega.$$

Again interchanging t and the dummy variable ω, and writing $T=\omega_0$, we obtain

$$\pi\{\delta(\omega-\omega_0)+\delta(\omega+\omega_0)\} = \int_{-\infty}^{\infty} \cos\omega_0 t\cdot e^{-j\omega t}\,dt.$$

It follows that the Fourier transform of the function $\cos\omega_0 t$ is

$$\pi\{\delta(\omega-\omega_0)+\delta(\omega+\omega_0)\}.$$

It is interesting to note how the *dual* nature of the Fourier transform pair allows the transforms of the functions 1 and $\cos\omega_0 t$ to be readily obtained.

These functions do not satisfy the condition

$$\int_{-\infty}^{\infty} |f(t)|\,dt < \infty,$$

and consequently their Fourier transforms contain non-elementary functions. Impulse functions are entirely theoretical, that is, they cannot be sketched or generated by physical equipment. The spectrum of $\cos \omega_0 t$ consists only of impulses 'of strength π' at $\omega = \pm \omega_0$. This confirms the obvious fact that all the *power* of the function $\cos \omega_0 t$ is concentrated at the frequency ω_0.

Problems

(1) Use a result from Example 1.5 to deduce that the Fourier transform of the function $\dfrac{\sin t}{t}$ is $\pi P_2(\omega)$.

(2) Find the Fourier transform of the function $f(t) = e^{-a|t|}, a > 0$, and deduce that

$$\int_0^{\infty} \frac{\cos \omega}{a^2 + \omega^2}\,d\omega = \frac{\pi}{2a} e^{-a}.$$

(3) Obtain the Fourier transform of the function

$$f(t) = \begin{cases} -e^{at} & \text{when } t < 0 \\ e^{-at} & \text{when } t > 0 \end{cases}, \; a > 0.$$

Hence show that this Fourier transform $\rightarrow \dfrac{2}{j\omega}$ as $a \rightarrow 0$, and deduce that the Fourier transform of the *unit step function u(t)*,

$$u(t) = \begin{cases} 1 & t \geqslant 0 \\ 0 & t < 0 \end{cases},$$

is given by $\pi \delta(\omega) + 1/j\omega$.

Answers to problems

(2) $\dfrac{2a}{a^2 + \omega^2}$.

(3) $\dfrac{1}{a + j\omega} - \dfrac{1}{a - j\omega}$.

1.5 THE LAPLACE TRANSFORM

In section 1.4, we saw the difficulties which arose when we attempted to obtain the Fourier transform of certain functions $f(t)$ which did not tend to zero as

$t \to \infty$. For these functions the condition

$$\int_{-\infty}^{\infty} |f(t)| \, dt < \infty$$

is not satisfied, and the integral

$$\int_{-\infty}^{\infty} f(t) e^{-j\omega t} \, dt$$

does not converge. The Laplace transform can be considered as a modification of the Fourier transform to overcome the difficulties described above. If we multiply the function $f(t)$ by $e^{-\sigma|t|}$ with $\sigma > 0$, then for a large class of functions the integral

$$\int_{-\infty}^{\infty} f(t) e^{-\sigma|t|} e^{-j\omega t} \, dt$$

will converge for a suitable value of σ. Most physical problems can be defined so that $f(t)$ is zero for $t < 0$, and the modified integral can be written

$$\int_{0}^{\infty} f(t) e^{-(\sigma + j\omega)t} \, dt.$$

Using the notation of equation (1.10), we have

$$F(\sigma + j\omega) = \int_{0}^{\infty} f(t) e^{-(\sigma + j\omega)t} \, dt. \tag{1.11}$$

Defining the variable $s = \sigma + j\omega$, we obtain

$$F(s) = \int_{0}^{\infty} f(t) e^{-st} \, dt. \tag{1.12}$$

$F(s)$ is called the **Laplace transform** of $f(t)$, and the complex variable s, whose imaginary part is the frequency variable, is known as the **Laplace variable**. Note that the Laplace transform of $f(t)$ is also the Fourier transform of $f(t) e^{-\sigma t}$, assuming that $f(t)$ is zero for $t < 0$.

Alternative notations for the transform pair are

$$F(s) = L\{f(t)\}, \quad f(t) = L^{-1}\{F(s)\}$$

where L is the Laplace transform *operator* and L^{-1} is the *inverse operator*. An inversion formula for the Laplace transform exists but involves an integration in the complex s-plane, and it is more convenient to use tabulated transform pairs for the common functions, see Table 1.

Table 1   Laplace transforms of $f(t)$, $f(t) = 0$ for $t < 0$.

$$L\{f(t)\} = \int_0^\infty f(t)\mathrm{e}^{-st}\,\mathrm{d}t = F(s)$$

$f(t)$	$F(s)$
1	$\dfrac{1}{s}$
t	$\dfrac{1}{s^2}$
t^2	$\dfrac{2}{s^3}$
$\dfrac{t^{n-1}}{(n-1)!}$, $n = 1, 2, 3, \ldots$	$\dfrac{1}{s^n}$
$tf(t)$	$-F'(s)$
$u(t-T)$	$\dfrac{\mathrm{e}^{-sT}}{s}$
$\delta(t)$	1
$\delta(t-T)$	e^{-sT}
$f(t-T)u(t-T)$	$\mathrm{e}^{-sT}F(s)$
e^{-at}	$\dfrac{1}{(s+a)}$
$t\mathrm{e}^{-at}$	$\dfrac{1}{(s+a)^2}$
$\cos\omega t$	$\dfrac{s}{s^2+\omega^2}$
$\sin\omega t$	$\dfrac{\omega}{s^2+\omega^2}$
$\mathrm{e}^{-at}\cos\omega t$	$\dfrac{(s+a)}{(s+a)^2+\omega^2}$
$\mathrm{e}^{-at}\sin\omega t$	$\dfrac{\omega}{(s+a)^2+\omega^2}$
$\mathrm{e}^{-at}f(t)$	$F(s+a)$
$f'(t)$	$sF(s)-f(0)$
$f''(t)$	$s^2F(s)-sf(0)-f'(0)$

Table 1   Continued.

$f(t)$	$F(s)$
$f^{(n)}(t)$	$s^n F(s) - \sum\limits_{r=1}^{n} s^{n-r} f^{(r-1)}(0)$
$\displaystyle\int_0^t f(u)\,du$	$\dfrac{F(s)}{s}$

Note: (i) In each case it is assumed that s is such that the integral
converges.

(ii) $f^{(n)}(t) \equiv \dfrac{d^n f}{dt^n}$.

Example 1.6

Obtain the Laplace transform of (i) the unit step function $u(t)$, (ii) the function
e^{-at}, $t \geqslant 0$, $a > 0$.

____

(i) $u(t) = \begin{cases} 1 & \text{when } t \geqslant 0 \\ 0 & \text{when } t < 0 \end{cases}$.

$$L\{u(t)\} = \int_0^{\infty} e^{-st}\,dt$$

$$= -\frac{1}{s}\left[e^{-st}\right]_0^{\infty}$$

$$= \frac{1}{s}, \qquad \text{assuming } \sigma > 0.$$

(ii) $L\{e^{-at}\} = \displaystyle\int_0^{\infty} e^{-(a+s)t}\,dt$

$$= -\frac{1}{s+a}\left[e^{-(a+s)t}\right]_0^{\infty}$$

$$= \frac{1}{s+a}, \qquad \text{assuming } \sigma > -a.$$

Example 1.7

(i) Writing $e^{j\omega t} = \cos \omega t + j \sin \omega t$, find the Laplace transforms of $\cos \omega t$ and
$\sin \omega t$.

(ii) Obtain the inverse Laplace transform of $\dfrac{2-3s}{(s+2)(s^2+4)}$.

(i) Following the working of part (ii) in Example 1.6,

$$L\{e^{j\omega t}\} = \frac{1}{s-j\omega} \qquad \text{assuming } \sigma > 0.$$

Taking real and imaginary parts,

$$L\{\cos \omega t\} = \frac{s}{s^2+\omega^2}$$

$$L\{\sin \omega t\} = \frac{\omega}{s^2+\omega^2}.$$

(ii) Using the method of partial fractions

$$\frac{2-3s}{(s+2)(s^2+4)} = \frac{1}{(s+2)} - \frac{s+1}{(s^2+4)}$$

$$= \frac{1}{(s+2)} - \frac{s}{(s^2+2^2)} - \frac{1}{2}\left(\frac{2}{s^2+2^2}\right).$$

$$\text{Therefore } L^{-1}\left\{\frac{2-3s}{(s+2)(s^2+4)}\right\} = e^{-2t} - \cos 2t - \tfrac{1}{2}\sin 2t,$$

using the results of part (i) and the previous example.

Example 1.8

Show that $L\left\{\dfrac{df}{dt}\right\} = sL\{f(t)\} - f(0)$.

$$L\left\{\frac{df}{dt}\right\} = \int_0^\infty \frac{df}{dt} e^{-st}\,dt$$

$$= \left[e^{-st}f(t)\right]_0^\infty + s\int_0^\infty f(t)e^{-st}\,dt,$$

(integration by parts).

$$\text{Therefore } L\left\{\frac{df}{dt}\right\} = -f(0) + sL\{f(t)\},$$

assuming that σ is large enough for $\lim_{t\to\infty} (e^{-st}f(t)) = 0$.

Problems

(1) Obtain the Laplace transform of the function

$$u(t-1) = \begin{cases} 1 & \text{when } t \geqslant 1 \\ 0 & \text{when } t < 1 \end{cases},$$

and hence write down the Laplace transform of the function

$$p(t) = \begin{cases} 1 & \text{when } 0 \leqslant t < 1 \\ 0 & \text{otherwise} \end{cases}$$

(2) Use Table 1 to obtain

(i) $\quad L^{-1}\left\{\dfrac{(s-3)^2}{s(s^2+9)}\right\}$ (ii) $\quad L^{-1}\left\{\dfrac{1-s-3s^2}{s^2(s+1)}\right\}.$

(3) Following the working of Example 1.8, show that the Fourier transform of df/dt is $j\omega F(j\omega)$, assuming that $f(t) \to 0$ as $t \to \pm\infty$.

Answers to problems

(1) $\dfrac{e^{-s}}{s}, \quad \dfrac{1-e^{-s}}{s}$

(2) (i) $1 - 2\sin 3t$ (ii) $t - 2 - e^{-t}.$

1.6 TRANSFER FUNCTIONS

In the introduction to this chapter we defined a continuous system as one which can be modelled by an ordinary differential equation. If we restrict our attention to systems described by *linear* differential equations with *constant coefficients*, then the concept of a transfer function becomes useful. For example, consider the system described by the differential equation

$$\frac{d^2y}{dt^2} + 3\frac{dy}{dt} + 2y = 3\frac{dx}{dt} + 2x. \tag{1.13}$$

where $x = x(t)$ and $y = y(t)$ are respectively the input and output of the system. From Problem 3 in section 1.5 the Fourier transform of $\dfrac{df}{dt}$ is $j\omega\, F(j\omega)$, and extending this result the transform of $\dfrac{d^2f}{dt^2}$ is $(j\omega)^2\, F(j\omega)$. Therefore Fourier transforming equation (1.13) gives

$$(j\omega)^2 Y(j\omega) + 3(j\omega)Y(j\omega) + 2Y(j\omega) = 3(j\omega)X(j\omega) + 2X(j\omega)$$

where $X(j\omega)$ and $Y(j\omega)$ are the Fourier transforms of $x(t)$ and $y(t)$.

Rearranging this transformed equation,

$$\frac{Y(j\omega)}{X(j\omega)} = \frac{3(j\omega) + 2}{(j\omega)^2 + 3(j\omega) + 2} = G(j\omega), \text{ say.}$$

$G(j\omega)$ is called the **transfer function** of the system. Given the input $x(t)$, and assuming that $X(j\omega)$ exists, $Y(j\omega)$ can be calculated by simply multiplying by the system transfer function. That is,

$$Y(j\omega) = G(j\omega)\, X(j\omega). \tag{1.14}$$

The system output $y(t)$ (and the solution of the differential equation) can now be obtained from the inversion formula.

For a system whose input 'starts' at $t = 0$ and whose output and its derivatives have zero values at $t = 0$ it is more convenient to use Laplace transforms, since these exist for all normal input functions.

From Example 1.8 $\quad L\left\{\dfrac{dy}{dt}\right\} = sL\{y\}$

when $y(0) = 0$, and this is easily extended to $\quad L\left\{\dfrac{d^2 y}{dt^2}\right\} = s^2 L\{y\}$

when $y(0) = \dot{y}(0) = 0$.

Taking Laplace transforms of equation (1.13) leads to

$$L\{y\} = G(s)\,L\{x\}. \tag{1.15}$$

where the *transfer function* $G(s)$ is the same function as $G(j\omega)$ above with $j\omega$ replaced by s. That is,

$$G(s) = \frac{3s + 2}{s^2 + 3s + 2}.$$

The transfer function of a system relates the *transformed* output and input. $G(s)$ is clearly independent of the input and is a valuable method of representing a linear continuous system.

For a given input function the output of the system can be obtained by taking inverse Laplace transforms of equation (1.15). In Chapters 2 and 3 the applications of transfer functions will be studied in more detail.

Closely related to the idea of a transfer function is the *convolution integral.* We define the **convolution** of two functions $x(t)$ and $g(t)$ as

$$x(t) * g(t) = \int_0^t x(u)\, g(t - u)\,du,$$

where it is assumed that both $x(t)$ and $g(t)$ are zero when $t < 0$. By considering $x(t)$ as a summation of narrow width delayed pulse functions it is easy to show that the solution of the equation (1.13) is given by

$$y(t) = x(t) * g(t). \tag{1.16}$$

where $g(t)$ is the solution to the differential equation for the special case $x(t) = \delta(t)$, the unit impulse function as defined in Example 1.5. However, we can approach this result by working in the transformed domain (or frequency domain). Following similar working to that given in Example 1.5 we can show that $L\{\delta(t)\} = 1$. For the special case $x(t) = \delta(t)$ equation (1.15) gives $L\{y\} = G(s)$, that is, $y(t) = L^{-1}\{G(s)\} = g(t)$, the *inverse Laplace transform of the system transfer function*, referred to as the **impulse response function**. For general $x(t)$ the solution is given by equation (1.15) as $y(t) = L^{-1}\{G(s)X(s)\}$, where $X(s)$ denotes $L\{x\}$. To prove equation (1.16) we require that

$$x(t) * g(t) = L^{-1}\{G(s)\,X(s)\}. \tag{1.17}$$

Equation (1.17) follows immediately from Theorem 1.1.

Theorem 1.1 (Convolution theorem)

$$L\{x(t) * g(t)\} = X(s)\,G(s).$$

Proof

$$L\{x(t) * g(t)\} = \int_0^\infty \left\{ \int_0^t x(u)\,g(t-u)\,\mathrm{d}u \right\} \mathrm{e}^{-st}\,\mathrm{d}t$$

$$= \int_0^\infty \int_0^\infty x(u)\,g(t-u)\mathrm{e}^{-st}\,\mathrm{d}u\,\mathrm{d}t, \text{ since } g(t-u) = 0 \text{ for } u > t.$$

Changing the order of integration of the repeated integral we obtain

$$L\{x(t) * g(t)\} = \int_0^\infty x(u) \int_0^\infty g(t-u)\mathrm{e}^{-st}\,\mathrm{d}t\,\mathrm{d}u$$

$$= \int_0^\infty x(u) \int_{-u}^\infty g(y)\mathrm{e}^{-s(y+u)}\,\mathrm{d}y\,\mathrm{d}u, \text{ where } y = t-u$$

$$= \int_0^\infty x(u) \int_0^\infty g(y)\mathrm{e}^{-sy}\mathrm{e}^{-su}\,\mathrm{d}y\,\mathrm{d}u, \text{ since } g(y) = 0 \text{ for } y < 0$$

$$= \int_0^\infty \mathrm{e}^{-su}x(u)\,\mathrm{d}u \cdot \int_0^\infty \mathrm{e}^{-sy}g(y)\,\mathrm{d}y$$

$$= X(s)\,G(s).$$

Thus we have a choice of methods for solving linear differential equations with constant coefficients.

(i) $y = \displaystyle\int_0^t x(u)\,g(t-u)\,\mathrm{d}u$, the **time domain** method,

(ii) $y = L^{-1}\{X(s)G(s)\}$ the **frequency domain** method.

The following theorem indicates a simple technique for obtaining the impulse response function.

Theorem 1.2
The differential equations

$$\ddot{y} + a\dot{y} + by = \delta(t), \quad y(0) = \dot{y}(0) = 0 \cdot \tag{i}$$

and

$$\ddot{y} + a\dot{y} + by = 0, \quad y(0) = 0, \quad \dot{y}(0) = 1 \tag{ii}$$

have identical solutions.

Proof
From Example 1.8 we have

$$L\{\dot{y}\} = sL\{y\} - y(0) = sL\{y\}.$$

Therefore $L\{\ddot{y}\} = sL\{\dot{y}\} - \dot{y}(0)$
$$= s^2 L\{y\} - sy(0) - \dot{y}(0).$$

Transforming equation (i) gives

$$s^2 L\{y\} + asL\{y\} + bL\{y\} = 1$$

Therefore $L\{y\} = \dfrac{1}{s^2 + as + b}.$

Transforming equation (ii) gives

$$s^2 L\{y\} - 1 + asL\{y\} + bL\{y\} = 0,$$

and $L\{y\} = \dfrac{1}{s^2 + as + b}$ as in (i).

(This theorem can be generalized to deal with nth order differential equations, where the impulse function on the right hand side is accommodated by imposing an initial condition on the $(n-1)$th derivative).

Example 1.9
Solve the differential equation,

$$\ddot{y} + 3\dot{y} + 2y = 2u(t), \quad y(0) = \dot{y}(0) = 0,$$

where $u(t)$ is the unit step function. Compare time domain analysis and frequency domain analysis.

(i) Time domain method
From Theorem 1.2 the impulse response function $g(t)$ is obtained by solving

$$\ddot{g} + 3\dot{g} + 2g = 0, \ g(0) = 0, \ \dot{g}(0) = 1, \text{ that is,}$$

$$g(t) = A\mathrm{e}^{-t} + B\mathrm{e}^{-2t}$$

where $A + B = 0$ and $-A - 2B = 1$.
Therefore $A = 1$, $B = -1$ and $g(t) = \mathrm{e}^{-t} - \mathrm{e}^{-2t}$.

$$\text{(Alternatively } g(t) = L^{-1}\left\{\frac{1}{s^2 + 3s + 2}\right\}$$

$$= L^{-1}\left\{\frac{1}{s+1} - \frac{1}{s+2}\right\}$$

$$= \mathrm{e}^{-t} - \mathrm{e}^{-2t} \qquad\qquad\qquad).$$

Equation (1.16) then gives

$$y(t) = \int_0^t 2(\mathrm{e}^{-(t-u)} - \mathrm{e}^{-2(t-u)})\,\mathrm{d}u$$

$$= 2\left[\mathrm{e}^{-(t-u)} - \frac{1}{2}\mathrm{e}^{-2(t-u)}\right]_0^t$$

$$= 2\left\{\frac{1}{2} - \mathrm{e}^{-t} + \frac{1}{2}\mathrm{e}^{-2t}\right\}.$$

Therefore the solution is

$$y = 1 - 2\mathrm{e}^{-t} + \mathrm{e}^{-2t}.$$

(ii) Frequency domain method
Taking Laplace transforms of the given differential equation,

$$(s^2 + 3s + 2)L\{y\} = \frac{2}{s}.$$

Therefore $L\{y\} = \dfrac{2}{s(s^2 + 3s + 2)}$

$$= \frac{1}{s} - \frac{2}{s+1} + \frac{1}{s+2}.$$

Using Table 1 to invert the transforms we obtain

$$y = 1 - 2\mathrm{e}^{-t} + \mathrm{e}^{-2t}.$$

Problems
(1) Show that $x(t)*g(t) = g(t)*x(t)$. Carry out the time domain solution in Example 1.9, using this alternative form of the convolution integral.

(2) Solve the differential equation

$$\frac{dy}{dt} + 3y = 13\cos 2t, \ y(0) = 0,$$

using (i) time domain analysis, (ii) frequency domain analysis.

(3) Solve equation (1.13) when $x(t) = u(t)$, $y(0) = \dot{y}(0) = 0$. Can you explain why your solution does *not* satisfy $\dot{y}(0) = 0$?

Answers to problems
(2) $y = -3e^{-3t} + 3\cos 2t + 2\sin 2t$.
(3) $y = 1 + e^{-t} - 2e^{-2t}$.
The definition of the impulse function implies that

$$\int_{-\infty}^{t} \delta(u)du = \begin{cases} 1 & \text{when } t > 0 \\ 0 & \text{when } t < 0 \end{cases} = u(t),$$

that is, $\dfrac{d}{dt}(u(t)) = \delta(t)$. The equation to be solved is

$$\ddot{y} + 3\dot{y} + 2y = 3\delta(t) + 2u(t), \ y(0) = \dot{y}(0) = 0.$$

Following Theorem 1.2 this is equivalent to

$$\ddot{y} + 3\dot{y} + 2y = 2, \ y(0) = 0, \ \dot{y}(0) = 3.$$

In fact $\dot{y}(t)$ has a discontinuity at $t = 0$,

that is, $\dot{y}(0^-) = 0$, $\dot{y}(0^+) = 3$.

CHAPTER 2

Solution of linear differential equations

2.1 INTRODUCTION

In example 1.9 we saw how the Laplace transform could be used to solve a certain linear differential equation with constant coefficients. The technique is formalized in this chapter, and the concept of a transfer function is again considered. Block diagram notation for continuous systems is introduced, and it is found convenient to describe systems in this form rather than use the differential equation model. For a given system input we shall see that many properties of the output can be obtained directly from the system transfer function without having to solve the corresponding differential equation.

2.2 LAPLACE TRANSFORMS OF COMMON FUNCTIONS

The Laplace transform of a function $f(t)$ defined for $t \geq 0$ is denoted by $F(s)$ and is defined as in equation (1.12),

$$F(s) = \int_0^\infty f(t)e^{-st}dt.$$

An alternative notation uses the *operator* L, that is

$$L\{f(t)\} = F(s).$$

The operator transforms a function of t into a function of s, and we saw in Chapter 1 that $s = \sigma + j\omega$ is a *complex* variable. For all useful functions $f(t)$ the Laplace transform will exist provided that the real part of s is sufficiently large, and in general we do not need to worry about convergence of the integral.

For a given function $F(s)$ we can carry out the inverse operation L^{-1}, to obtain $f(t) = L^{-1}\{F(s)\}$, using an inversion formula similar to equation (1.9). However, for the usual functions it is more convenient to use Table 1 which lists all common transform pairs. In section 1.5 we obtained the following Laplace transforms

$$L\{u(t)\} = \frac{1}{s},$$

$$L\{e^{-at}\} = \frac{1}{s+a},$$

$$L\{\cos\omega t\} = \frac{s}{s^2+\omega^2},$$

$$L\{\sin\omega t\} = \frac{\omega}{s^2+\omega^2}.$$

The following examples and theorems extend the list of transforms.

Example 2.1
Obtain the Laplace transform of the function $f(t) = t$.

$$\begin{aligned}
L\{t\} &= \int_0^\infty t\,e^{-st}\,dt \\[2mm]
&= \left[\frac{te^{-st}}{-s}\right]_0^\infty + \frac{1}{s}\int_0^\infty e^{-st}\,dt \\[2mm]
&= 0 - \frac{1}{s^2}\left[e^{-st}\right]_0^\infty \\[2mm]
&= \frac{1}{s^2}, \text{ assuming } \sigma > 0.
\end{aligned}$$

Theorem 2.1
Given that $L\{f(t)\} = F(s)$, then

$$L\{tf(t)\} = -\frac{d}{ds}F(s).$$

Proof

$$F(s) = \int_0^\infty f(t)e^{-st}\,dt.$$

Differentiating† with respect to s,

$$\begin{aligned}
F'(s) &= -\int_0^\infty tf(t)e^{-st}\,dt \\[2mm]
&= -L\{tf(t)\}.
\end{aligned}$$

Therefore $L\{tf(t)\} = -F'(s)$.

[† This step requires the uniform convergence of the integral

$$\int_0^\infty tf(t)e^{-st}\,dt, \text{ see Apostol (1957)}].$$

Example 2.2
Use Theorem 2.1 to obtain the Laplace transform of

(i) t^2 (ii) te^{-at}.

(i) $L\{t^2\} = L\{t \cdot t\} = -\dfrac{d}{ds}L\{t\}$

$$= -\frac{d}{ds}\left(\frac{1}{s}\right) = \frac{2}{s^3}$$

(ii) $L\{te^{-at}\} = -\dfrac{d}{ds}L\{e^{-at}\}$

$$= -\frac{d}{ds}\left(\frac{1}{s+a}\right)$$

$$= \frac{1}{(s+a)^2}.$$

Theorem 2.2
Given that $L\{f(t)\} = F(s)$, then

$$L\{e^{-at}f(t)\} = F(s+a).$$

Proof

$$L\{e^{-at}f(t)\} = \int_0^\infty e^{-at}f(t)e^{-st}\,dt$$

$$= \int_0^\infty f(t)e^{-(s+a)t}\,dt$$

$$= F(s+a), \text{ that is, } F(s) \text{ with } s \text{ replaced by } s+a.$$

Example 2.3
Use theorem 2.2 to write down the Laplace transforms of $e^{-at}\cos \omega t$ and $e^{-at}\sin \omega t$

$$L\{\cos \omega t\} = \frac{s}{s^2+\omega^2} \text{ and } L\{\sin \omega t\} = \frac{\omega}{s^2+\omega^2}.$$

Therefore $L\{e^{-at}\cos \omega t\} = \dfrac{s+a}{(s+a)^2+\omega^2},$

$$L\{e^{-at}\sin \omega t\} = \frac{\omega}{(s+a)^2+\omega^2}.$$

Example 2.4

Obtain the inverse Laplace transform of $\dfrac{\omega}{s(s^2+\omega^2)}$,

(i) using partial fractions and Table 1,
(ii) using the convolution theorem (Theorem 1.1)

(i) $\dfrac{\omega}{s(s^2+\omega^2)} = \dfrac{1}{\omega}\left(\dfrac{1}{s}-\dfrac{s}{s^2+\omega^2}\right).$

Therefore $L^{-1}\left\{\dfrac{\omega}{s(s^2+\omega^2)}\right\} = \dfrac{1}{\omega}(1-\cos\omega t).$

(ii) From Theorem 1.1

$$L^{-1}\{X(s)G(s)\} = \int_0^t x(u)g(t-u)\,du.$$

Let $X(s) = \dfrac{1}{s}, G(s) = \dfrac{\omega}{s^2+\omega^2}.$

Therefore $x(t) = 1, g(t) = \sin\omega t,$

and $L^{-1}\left\{\dfrac{1}{s}\cdot\dfrac{\omega}{s^2+\omega^2}\right\} = \int_0^t \sin\omega(t-u)\,du$

$$= \left[\dfrac{1}{\omega}\cos\omega(t-u)\right]_0^t = \dfrac{1}{\omega}(1-\cos\omega t).$$

Problems
(1) Obtain the Laplace transforms of the following functions
    (i) t^3 (i) $e^{-2t}\cos 3t$
    (iii) $t^2 e^{-3t}$ (iv) $\cosh 2t$
    (v) $\cos^2 t.$
(2) Find the inverse Laplace transforms of the following functions.
    (i) $\dfrac{2}{s(s-1)}$ (ii) $\dfrac{2}{(s+2)^2+4^2}$
    (iii) $\dfrac{2}{(s-1)(s^2+1)}.$

(3) Apply Theorem 2.1 to show that

$$L\{t^n\} = \dfrac{n!}{s^{n+1}}, \quad n \text{ a positive integer.}$$

(4) (a) Use the convolution theorem to show that

$$L^{-1}\left\{\frac{\omega}{s^2(s^2+\omega^2)}\right\} = \frac{\omega t - \sin\omega t}{\omega^2}.$$

(b) Find the inverse Laplace transform of

$$\frac{3s}{(s^2+1)(s^2+4)}.$$

Answers to problems

(1) (i) $\dfrac{6}{s^4}$ (ii) $\dfrac{s+2}{s^2+4s+13}$ (iii) $\dfrac{2}{(s+3)^3}$

     (iv) $\dfrac{s}{s^2-4}$ (v) $\dfrac{1}{2}\left(\dfrac{1}{s}+\dfrac{s}{s^2+4}\right)$

(2) (i) $-2+2e^t$ (ii) $\frac{1}{2}e^{-2t}\sin 4t$
     (iii) $e^t - \cos t - \sin t$
(4) (b) $\cos t - \cos 2t$.

2.3 SOLUTION OF INITIAL VALUE PROBLEMS

An nth order differential equation together with given *initial* values (values at $t = 0$) of the dependent variable and its derivatives up to the $(n-1)$th derivative constitutes an **initial value problem**. For example, the differential equation

$$\ddot{y}+3\dot{y}+2y = 1-t,$$

together with the conditions $y(0) = 1$, $\dot{y}(0) = -2$, represents a second order initial value problem.

   We restrict our attention to differential equations which are linear with constant coefficients. In this case the application of Laplace transforms results in an *algebraic* equation which can be solved for the Laplace transform of the dependent variable. The method depends on the following two theorems.

Theorem 2.3

$L\{af(t)+bg(t)\} = aL\{f(t)\} + bL\{g(t)\}$, where a, b are constants.

   The proof follows immediately from the definition of the Laplace transform. The theorem is referred to as the **linearity property** of Laplace transforms.

Theorem 2.4

$$L\{f^{(n)}(t)\} = s^n L\{f(t)\} - \sum_{r=1}^{n} s^{n-r}f^{(r-1)}(0),$$

where $f^{(n)}(t) \equiv \dfrac{\mathrm{d}^n f}{\mathrm{d}t^n}$ the nth derivative of $f(t)$.

Proof (By induction)

Example 1.8 proved the result for $n = 1$, that is,

$$L\left\{\frac{\mathrm{d}f}{\mathrm{d}t}\right\} = s\,L\{f(t)\} - f(0). \tag{2.1}$$

We assume the result for $n = k$, that is,

$$L\{f^{(k)}(t)\} = s^k\,L\{f(t)\} - \sum_{r=1}^{k} s^{k-r} f^{(r-1)}(0). \tag{2.2}$$

Now $L\{f^{(k+1)}(t)\} = L\left\{\dfrac{\mathrm{d}}{\mathrm{d}t} f^{(k)}(t)\right\}$

$$= s\,L\{f^{(k)}(t)\} - f^{(k)}(0),$$

using equation (2.1).

Therefore substituting from equation (2.2)

$$L\{f^{(k+1)}(t)\} = s^{k+1}\,L\{f(t)\} - \sum_{r=1}^{k} s^{k+1-r} f^{(r-1)}(0) - f^{(k)}(0)$$

$$= s^{k+1}\,L\{f(t)\} - \sum_{r=1}^{k+1} s^{k+1-r} f^{(r-1)}(0).$$

This last equation represents the result for $n = k+1$. Therefore we have proved that *if* the result is true for $n = k$, it is *also* true for $n = k+1$. Since we know that the result is true for $n = 1$, it follows that it is true for $n = 2, n = 3,$ ..., that is for all n.

Examples 2.5, 2.6, and 2.7 demonstrate the application of Laplace transforms to the solution of differential equations.

Example 2.5

Solve the initial value problem

$$\ddot{y} + 3\dot{y} + 2y = 1 - t, \quad y(0) = 1, \dot{y}(0) = -2$$

From Theorem 2.4

$$L\{\ddot{y}\} = s^2 L\{y\} - s y(0) - \dot{y}(0) = s^2 L\{y\} - s + 2,$$

$$L\{\dot{y}\} = s L\{y\} - y(0) = s L\{y\} - 1.$$

The linearity property (Theorem 2.3) allows us to transform the differential equation term by term.

$$s^2 L\{y\} - s + 2 + 3(sL\{y\} - 1) + 2L\{y\} = \frac{1}{s} - \frac{1}{s^2}$$

that is, $(s^2 + 3s + 2)L\{y\} = s + 1 + \dfrac{1}{s} - \dfrac{1}{s^2} = \dfrac{s^3 + s^2 + s - 1}{s^2}.$

$$\begin{aligned}
\text{Therefore } L\{y\} &= \frac{s^3 + s^2 + s - 1}{s^2(s^2 + 3s + 2)} \\[2mm]
&= \frac{s^3 + s^2 + s - 1}{s^2(s+1)(s+2)} \\[2mm]
&= \frac{5}{4s} - \frac{1}{2s^2} - \frac{2}{s+1} + \frac{7}{4(s+2)},
\end{aligned}$$

(partial fractions).

Using Table 1 to invert the Laplace transforms we obtain the required solution,

$$y = \frac{5}{4} - \frac{t}{2} - 2e^{-t} + \frac{7}{4}e^{-2t}.$$

Example 2.6

Solve the differential equation

$$\dddot{y} + 3\ddot{y} + 4\dot{y} + 2y = 10,$$

given the initial conditions $y(0) = \dot{y}(0) = \ddot{y}(0) = 0.$

Since all the initial conditions are zero, Theorem 2.4 gives

$$L\{\dddot{y}\} = s^3 L\{y\}, \quad L\{\ddot{y}\} = s^2 L\{y\}, \quad L\{\dot{y}\} = sL\{y\}.$$

Transforming the equation gives

$$(s^3 + 3s^2 + 4s + 2)L\{y\} = \frac{10}{s}.$$

$$\begin{aligned}
\text{Therefore } L\{y\} &= \frac{10}{s(s^3 + 3s^2 + 4s + 2)} \\[2mm]
&= \frac{10}{s(s+1)(s^2 + 2s + 2)} \\[2mm]
&= \frac{5}{s} - \frac{10}{s+1} + \frac{5s}{s^2 + 2s + 2} \\[2mm]
&= \frac{5}{s} - \frac{10}{s+1} + \frac{5(s+1)}{(s+1)^2 + 1} - \frac{5}{(s+1)^2 + 1}.
\end{aligned}$$

Note the use of 'completing the square' to obtain expressions which appears in Table 1. Inverting the Laplace transforms we obtain the solution,

$$y = 5 - 10e^{-t} + 5e^{-t}\cos t - 5e^{-t}\sin t.$$

Example 2.7
Solve the differential equation

$$\ddot{x} - x = \cos 2t,$$

subject to the conditions $x(0) = 0$, $x(\pi) = 0$.

This is not an *initial* value problem, since the value of $\dot{x}(0)$ is not given. However, we can assign an arbitrary value A for $\dot{x}(0)$ and subsequently evaluate A to satisfy the condition $x(\pi) = 0$.

Taking Laplace transforms

$$s^2 L\{x\} - A - L\{x\} = \frac{s}{s^2 + 4}.$$

Therefore $(s^2 - 1)L\{x\} = A + \dfrac{s}{s^2 + 4}$, that is,

$$L\{x\} = \frac{A}{(s-1)(s+1)} + \frac{s}{(s-1)(s+1)(s^2+4)}$$

$$= \frac{A}{2}\left(\frac{1}{(s-1)} - \frac{1}{(s+1)}\right) + \frac{1}{10}\left(\frac{1}{(s-1)} + \frac{1}{(s+1)} - \frac{2s}{(s^2+4)}\right).$$

Therefore $x = \dfrac{A}{2}(e^t - e^{-t}) + \dfrac{1}{10}(e^t + e^{-t}) - \dfrac{1}{5}\cos 2t.$

We have to satisfy $x(\pi) = 0$, that is,

$$0 = \frac{A}{2}(e^\pi - e^{-\pi}) + \frac{1}{10}(e^\pi + e^{-\pi}) - \frac{1}{5}.$$

Therefore $A = \dfrac{2 - e^\pi - e^{-\pi}}{5(e^\pi - e^{-\pi})} \simeq -0.1834$

and the solution is approximately

$$x = 0.0083e^t + 0.1917e^{-t} - 0.2\cos 2t.$$

Example 2.8
Show that the Laplace transform of $\displaystyle\int_0^t f(u)\,du$ is $\dfrac{1}{s}L\{f(t)\}.$

Define $g(t) = \int_0^t f(u)\,du$ and note that

$$\dot{g}(t) = f(t), \quad g(0) = 0.$$

Applying Theorem 2.4,

$$L\{\dot{g}(t)\} = sL\{g(t)\} - g(0),$$

that is,

$$L\{f(t)\} = sL\{g(t)\}.$$

Therefore

$$L\{g(t)\} = \frac{1}{s}L\{f(t)\}, \text{ which proves the result.}$$

It is interesting to note that the Laplace transform of an integral involves a division by s, and the Laplace transform of a derivative involves a multiplication by s. This demonstrates a correspondence between the Laplace variable s and the differential operator $D \equiv d/dt$.

Problems
(1) Solve the following second order initial value problems.
    (i) $\ddot{y} + 2\dot{y} + 2y = 10\sin 2t$, $y(0) = -1$, $\dot{y}(0) = -3$.
    (ii) $\ddot{y} + \dot{y} - 2y = \sin t$, $y(0) = 1$, $\dot{y}(0) = -1$.
    (iii) $\ddot{x} - 2\dot{x} - 3x = 3t^2 + 7t + 3$, $x(0) = \dot{x}(0) = -1$.
    (iv) $\ddot{y} + 2\dot{y} + 2y = 5\sin t$, $y(0) = \dot{y}(0) = 0$.
    (v) $\ddot{x} - \dot{x} - 2x = 3e^{2t}$, $x(0) = 0$, $\dot{x}(0) = 2$.
(2) Solve the following initial value problems.
    (i) $\dddot{y} - 3\ddot{y} + 2\dot{y} = 3e^t$, $y(0) = 0$, $\dot{y}(0) = 1$, $\ddot{y}(0) = 2$.
    (ii) $\dddot{x} - 2\ddot{x} + \dot{x} = 2$, $x(0) = 2$, $\dot{x}(0) = 1$, $\ddot{x}(0) = -1$.

    (iii) $\dfrac{d^4\theta}{dt^4} - \theta = 0$, $\theta(0) = 1$, $\dot{\theta}(0) = \ddot{\theta}(0) = \dddot{\theta}(0) = 0$.

(3) Use Laplace transforms to find the *general* solution of the following differential equations.
    (i) $\ddot{y} - 8y = 16$.
    (ii) $\dddot{y} - 2\ddot{y} + 2\dot{y} - y = 0$.
(4) Use Laplace transforms to solve the *boundary* value problems.
    (i) $\ddot{y} + 0.25y = 0$, $y(0) = 0$, $y(\pi) = 0.5$.
    (ii) $\ddot{x} + 2\dot{x} = 4t$, $x(0) = x(1) = 0$.

(5) Prove that $L\left\{\dfrac{f(t)}{t}\right\} = \displaystyle\int_s^\infty F(u)\,du$,

provided that the integral exists, where $F(s) = L\{f(t)\}$. Hence show that

$$L\left\{\frac{\sin t}{t}\right\} = \tan^{-1}\left(\frac{1}{s}\right).$$

Answers to problems
(1) (i) $y = e^{-t}\cos t - 2\cos 2t - \sin 2t$.
    (ii) $y = 0.5e^{t} + 0.6e^{-2t} - 0.1\cos t - 0.3\sin t$.
    (iii) $x = -1 - t - t^{2}$.
    (iv) $y = \sin t - 2\cos t + e^{-t}(\sin t + 2\cos t)$.
    (v) $x = \frac{1}{3}(e^{2t} - e^{-t}) + te^{2t}$.
(2) (i) $y = \frac{1}{9}(e^{t} - e^{-2t}) + \frac{2}{3}te^{t} + \frac{1}{2}t^{2}e^{t}$.
    (ii) $x = 3 + 2t - e^{t}$
    (iii) $\theta = 0.25(e^{t} + e^{-t}) + 0.5\cos t$.
(3) (i) $y = -2 + c_{1}e^{2t} + e^{-t}(c_{2}\cos\sqrt{3}t + c_{3}\sin\sqrt{3}t)$.
    (ii) $y = c_{1}e^{-t} + (c_{2} + c_{3}t + c_{4}t^{2})e^{t}$,
       where $c_{1}, c_{2}, c_{3}, c_{4}$, are arbitrary constants.
(4) (i) $y = 0.5\sin 0.5t$.
    (ii) $x = -t + t^{2}$.

2.4 SIMULTANEOUS DIFFERENTIAL EQUATIONS

Continuous systems with multiple inputs and outputs may be represented by systems of simultaneous differential equations. Provided these differential equations are linear with constant coefficients and the inputs belong to a certain class of functions, then they can be expressed in *first order homogeneous form*, $\dot{\mathbf{x}} = A\mathbf{x}$ where $\mathbf{x}$ is an n-dimensional vector and A is an $n \times n$ matrix with constant elements.

Huntley & Johnson (1983) consider a time domain method of solution which depends on algebraic properties of the matrix. Alternatively the solution may be obtained by taking Laplace transforms of $\dot{\mathbf{x}} = A\mathbf{x}$ to obtain

$$sL\{\mathbf{x}\} - \mathbf{x}(0) = AL\{\mathbf{x}\}. \tag{2.3}$$

Equation (2.3) can be arranged to give

$$(sI - A)L\{\mathbf{x}\} = \mathbf{x}(0).$$

Therefore $$L\{\mathbf{x}\} = (sI - A)^{-1}\mathbf{x}(0); \tag{2.4}$$

and the solution of the system of differential equations is obtained by taking inverse Laplace transforms of equation (2.4). Example 2.9 demonstrates the method. It should be noted that the matrix inverse $(sI - A)^{-1}$ involves division by $|sI - A|$, the determinant of the matrix $sI - A$. If $x = a$ is a solution of the polynomial equation $|sI - A| = 0$ then the solution of the differential equations will include terms e^{at}. The set of solutions of $|sI - A| = 0$ are, of course, the *eigenvalues* of the matrix A, see Huntley & Johnson (1983).

The advantage of Laplace transform techniques applied to simultaneous differential equations is that it is *not necessary* to reduce the equations to the form $\dot{\mathbf{x}} = A\mathbf{x}$, see examples 2.10 and 2.11.

Example 2.9

Solve the system of differential equations

$$\dot{x}_1 = x_1 - 2x_2$$
$$\dot{x}_2 = 5x_1 + 3x_2$$

given the conditions $x_1(0) = 1$, $x_2(0) = 1$.

Writing the system in matrix form,

$$\begin{bmatrix} \dot{x}_1 \\ \dot{x}_2 \end{bmatrix} = \begin{bmatrix} 1 & -2 \\ 5 & 3 \end{bmatrix} \begin{bmatrix} x_1 \\ x_2 \end{bmatrix}, \begin{bmatrix} x_1(0) \\ x_2(0) \end{bmatrix} = \begin{bmatrix} 1 \\ 1 \end{bmatrix}.$$

(This step is not strictly necessary; it is easy to transform the equations as they stand and solve for $L\{x_1\}$ and $L\{x_2\}$).

Taking Laplace transforms of the matrix equation gives

$$s\begin{bmatrix} L\{x_1\} \\ L\{x_2\} \end{bmatrix} - \begin{bmatrix} 1 \\ 1 \end{bmatrix} = \begin{bmatrix} 1 & -2 \\ 5 & 3 \end{bmatrix} \begin{bmatrix} L\{x_1\} \\ L\{x_2\} \end{bmatrix},$$

that is,

$$\begin{bmatrix} (s-1) & 2 \\ -5 & (s-3) \end{bmatrix} \begin{bmatrix} L\{x_1\} \\ L\{x_2\} \end{bmatrix} = \begin{bmatrix} 1 \\ 1 \end{bmatrix}.$$

Therefore

$$\begin{bmatrix} L\{x_1\} \\ L\{x_2\} \end{bmatrix} = \frac{1}{(s^2 - 4s + 13)} \begin{bmatrix} (s-3) & -2 \\ 5 & (s-1) \end{bmatrix} \begin{bmatrix} 1 \\ 1 \end{bmatrix},$$

that is,

$$L\{x_1\} = \frac{s-5}{s^2 - 4s + 13} = \frac{s-2}{(s-2)^2 + 3^2} - \frac{3}{(s-2)^2 + 3^2}$$

and

$$L\{x_2\} = \frac{s+4}{s^2 - 4s + 13} = \frac{s-2}{(s-2)^2 + 3^2} + 2\frac{3}{(s-2)^2 + 3^2}.$$

Inverting the transforms gives the solution

$$x_1 = e^{2t}\cos 3t - e^{2t}\sin 3t$$
$$x_2 = e^{2t}\cos 3t + 2e^{2t}\sin 3t.$$

Example 2.10

Solve the simultaneous differential equations

$$\ddot{x} = 2x + \dot{x} + y$$
$$\dot{y} = 4x + 2y + 1$$

given that $x(0) = 1$, $\dot{x}(0) = 0$, $y(0) = 0$.

Taking Laplace transforms of the given equations,

$$s^2 L\{x\} - s = 2L\{x\} + sL\{x\} - 1 + L\{y\}$$

$$sL\{y\} = 4L\{x\} + 2L\{y\} + \frac{1}{s},$$

that is, $(s^2 - s - 2)L\{x\} - L\{y\} = s - 1$

$$-4L\{x\} + (s - 2)L\{y\} = \frac{1}{s}.$$

Solving these *algebraic* equations for $L\{x\}$ gives

$$L\{x\} = \frac{(s-2)(s-1) + \dfrac{1}{s}}{(s^2 - s - 2)(s - 2) - 4}$$

$$= \frac{s^3 - 3s^2 + 2s + 1}{s^3(s - 3)}$$

$$= \frac{20}{27s} - \frac{7}{9s^2} - \frac{1}{3s^3} + \frac{7}{27(s - 3)}.$$

Therefore $x = \dfrac{20}{27} - \dfrac{7t}{9} - \dfrac{t^2}{6} + \dfrac{7e^{3t}}{27}.$

y can be obtained by solving the algebraic equations for $L\{y\}$, or alternatively by substituting for x in the first differential equation. That is,

$$y = \ddot{x} - \dot{x} - 2x$$

$$= -\frac{1}{3} + \frac{7e^{3t}}{3} - \left(-\frac{7}{9} - \frac{t}{3} + \frac{7e^{3t}}{9}\right) - \left(\frac{40}{27} - \frac{14t}{9} - \frac{t^2}{3} + \frac{14e^{3t}}{27}\right).$$

Therefore $y = -\dfrac{28}{27} + \dfrac{17t}{9} + \dfrac{t^2}{3} + \dfrac{28e^{3t}}{27}.$

Example 2.11

Consider the undamped mechanical system shown in Fig. 2.1. Assuming linear springs and neglecting friction in the wheels, the equations of motion are

$$m_1 \ddot{y}_1 = k_2(y_2 - y_1) - k_1 y_1$$

$$m_2 \ddot{y}_2 = -k_2(y_2 - y_1).$$

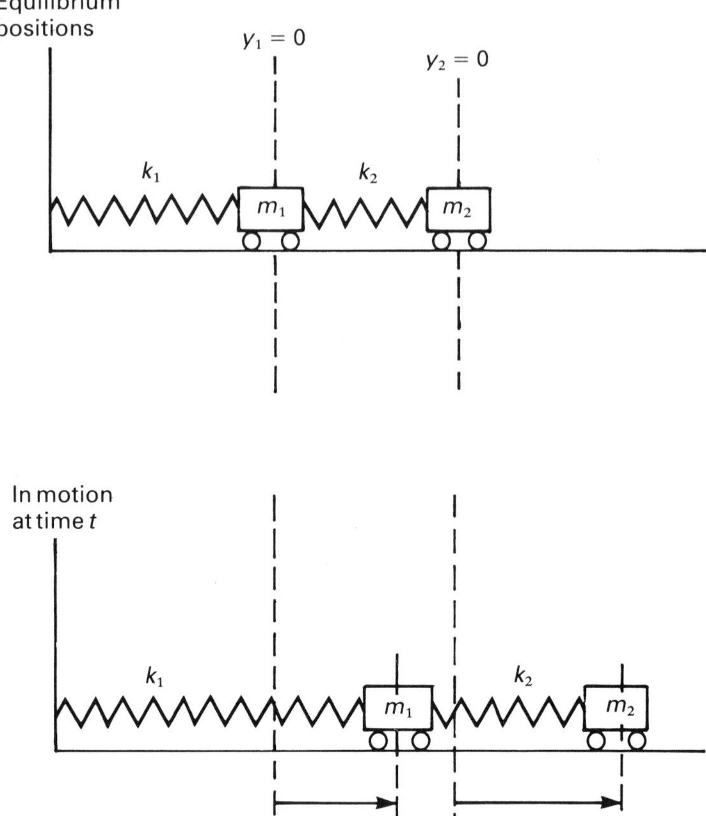

Fig. 2.1 k_1, k_2 are the spring stiffness; m_1, m_2 are the masses of the trucks.

Solve the equations when $\dfrac{k_1}{m_1} = 12$,

$$\frac{k_2}{m_1} = \frac{k_2}{m_2} = 8, \text{ given the initial conditions}$$

$$y_1(0) = y_2(0) = \dot{y}_1(0) = 0, \ \dot{y}_2(0) = 24.$$

Using the given numbers the equations of motion are

$$\ddot{y}_1 = -20y_1 + 8y_2$$
$$\ddot{y}_2 = 8y_1 - 8y_2.$$

Taking Laplace transforms we obtain

$$s^2 L\{y_1\} = -20L\{y_1\} + 8L\{y_2\}$$
$$s^2 L\{y_2\} - 24 = 8L\{y_1\} - 8L\{y_2\},$$

that is,

$$(s^2 + 20)L\{y_1\} - 8L\{y_2\} = 0$$
$$-8L\{y_1\} + (s^2 + 8)L\{y_2\} = 24.$$

Solving for $L\{y_1\}$,

$$L\{y_1\} = \frac{192}{(s^2 + 20)(s^2 + 8) - 64}$$

$$= \frac{192}{(s^2 + 4)(s^2 + 24)}$$

$$= 4.8\left(\frac{2}{s^2 + 2^2}\right) - \frac{9.6}{\sqrt{24}}\left(\frac{\sqrt{24}}{s^2 + (\sqrt{24})^2}\right).$$

Therefore $y_1 = 4.8 \sin 2t - 1.96 \sin \sqrt{24}t$, approximately.

From the first equation of motion

$$y_2 = \frac{1}{8}(\ddot{y}_1 + 20y_1)$$

$$= \frac{1}{8}(-19.2 \sin 2t + 47.04 \sin \sqrt{24}t + 96 \sin 2t - 39.2 \sin \sqrt{24}t).$$

Therefore $y_2 = 9.6 \sin 2t + 0.98 \sin \sqrt{24}t$, approximately.

Problems

(1) Solve the following systems of first order differential equations for the given initial conditions.

(i) $\dot{x}_1 = 16x_1 - 8x_2$
$\dot{x}_2 = -12x_1 + 12x_2$, $x_1(0) = 2$, $x_2(0) = 3$.

(ii)

$$\begin{bmatrix} \dot{x}_1 \\ \dot{x}_2 \\ \dot{x}_3 \end{bmatrix} = \begin{bmatrix} 0 & 1 & 1 \\ 0 & -1 & 1 \\ 0 & 0 & -1 \end{bmatrix} \begin{bmatrix} x_1 \\ x_2 \\ x_3 \end{bmatrix}, \quad \begin{bmatrix} x_1(0) \\ x_2(0) \\ x_3(0) \end{bmatrix} = \begin{bmatrix} 2 \\ 0 \\ -1 \end{bmatrix}.$$

(iii) $\dot{y}_1 = 0.6y_1 + 0.8y_2 - 1$
$\dot{y}_2 = 0.8y_1 - 0.6y_2 + 2$
$y_1(0) = 0$, $y_2(0) = 5$.

(iv) $\dot{x} = y + \sin t$
$\dot{y} = 2x + y + 2\cos t$
$x(0) = 0$, $y(0) = -1$.

(2) Solve the following initial value problems.

   (i) $\ddot{y} + 6\dot{y} + 8y = 8x$.

      $\dot{x} + 2x = 0$, $x(0) = 2$, $y(0) = 1$, $\dot{y}(0) = 4$.

   (ii) $\dot{x}_1 = x_2$

      $\ddot{x}_2 = -\dot{x}_1 + 2\dot{x}_2$, $x_1(0) = 1$, $x_2(0) = 1$, $\dot{x}_2(0) = 2$.

   (iii) $\ddot{x}_1 = -10x_1 + 6x_2 + 26e^{-t}$

      $\ddot{x}_2 = 15x_1 - 19x_2$

      $x_1(0) = 4$, $x_2(0) = 3$, $\dot{x}_1(0) = 0$, $\dot{x}_2(0) = 1$.

(3) Fig. 2.2 shows a mechanical system in motion at time t; the positions of the trucks are measured from the equilibrium positions. The equations of motion are

$$m_1\ddot{y}_1 = -(k_1 + k_2)y_1 + k_2 y_2$$
$$m_2\ddot{y}_2 = k_2 y_1 - (k_2 + k_3)y_2.$$

Given that $k_1 = k_2 = k_3 = m_1 = m_2 = 1$,

$y_1(0) = y_2(0) = \dot{y}_1(0) = 0$, $\dot{y}_2(0) = 6$,

express y_1 and y_2 in terms of t.

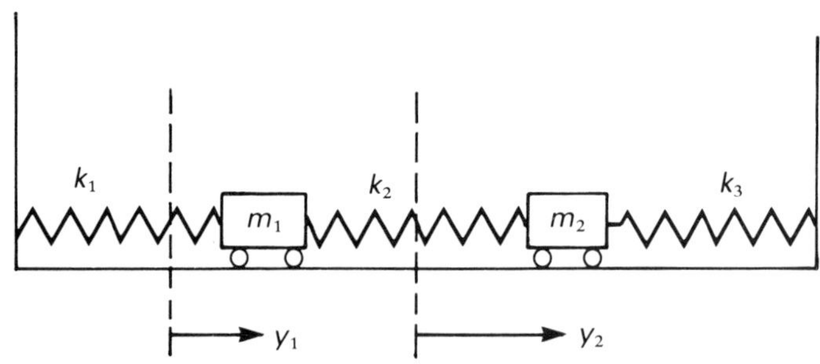

Fig. 2.2 k_1, k_2, k_3 are the spring stiffness; m_1, m_2 are the masses of the trucks.

(4) An undamped mechanical system with n degrees of freedom is represented by the system of second order differential equations $\ddot{\mathbf{y}} = A\mathbf{y}$, where $\mathbf{y}$ is an n-dimensional vector. Show that, in general, the solution contains terms of frequency $\sqrt{-\lambda}$ where λ is a solution of the polynomial equation $|\lambda I - A| = 0$.

Answers to problems

(1) (i) $x_1 = 2e^{4t}$, $x_2 = 3e^{4t}$.

    (ii) $\begin{bmatrix} x_1 \\ x_2 \\ x_3 \end{bmatrix} = \begin{bmatrix} 2+t \\ -t \\ -1 \end{bmatrix} e^{-t}$.

$\qquad$ (iii) $y_1 = 2e^t - e^{-t} - 1.$
$\qquad\qquad y_2 = e^t + 2e^{-t} + 2.$
$\qquad$ (iv) $x = e^{-t} - \cos t,\ y = -e^{-t}.$
(2) (i) $\quad x = 2e^{-2t},\ y = e^{-4t} + 8te^{-2t}.$
$\qquad$ (ii) $x_1 = 1 + te^t,\ x_2 = e^t + te^t.$
$\qquad$ (iii) $x_1 = 2\sin 2t + 4e^{-t}$
$\qquad\qquad x_2 = 2\sin 2t + 3e^{-t}.$
(3) $y_1 = 3\sin t - \sqrt{3}\sin\sqrt{3}t$
$\qquad y_2 = 3\sin t + \sqrt{3}\sin\sqrt{3}t.$

2.5 BLOCK DIAGRAM NOTATION

We have seen in section 1.6 that a continuous system with a single input and single output (a SISO system) can be represented either by a differential equation or in certain cases by a transfer function. The differential equation relates the input $x(t)$ to the output $y(t)$, the latter being given by the convolution integral $y(t) = \int_0^t x(u)g(t-u)\,du$, where $g(t)$ is the impulse response function. The transfer function $G(s) = L\{g(t)\}$ is a particularly convenient way of representing the system since the transformed output is related to the transformed input by simple multiplication, $L\{y(t)\} = L\{x(t)\}\cdot G(s)$. We refer to the system with transfer function $G(s)$ as simply 'the system $G(s)$' and use the convenient **block diagram notation** shown in Fig. 2.3 to represent the system.

It is a simple matter to convert a linear differential equation with constant coefficients to block diagram notation by taking Laplace transforms, assuming that the input, output, and their derivatives have zero values when $t = 0$. The reverse process of isolating the differential equation given the block diagram notation is achieved by replacing the Laplace parameter s by the differential operator $D \equiv d/dt$. Examples 2.12 and 2.13 demonstrate the techniques.

If the output from a system $G_1(s)$ is made the input to a second system $G_2(s)$ then the output of the second system is related to the input of the first system by the transfer function $G_1(s)G_2(s)$. In other words the overall transfer function of a number of systems in *series* is obtained by multiplying the individual transfer functions. Similarly, systems in *parallel* can be combined

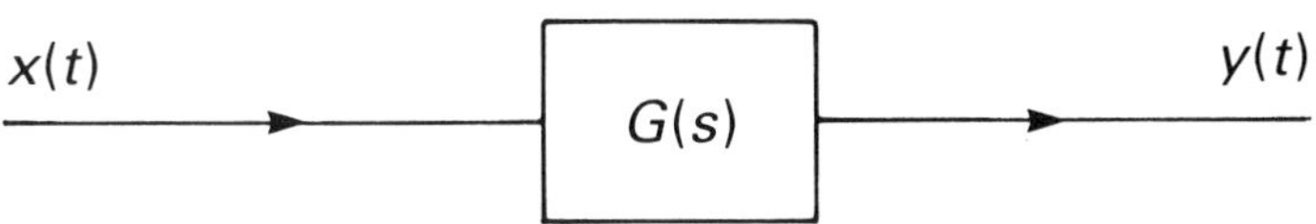

Fig. 2.3 Block diagram representation of system with transfer function $G(s)$, that is, $L\{y(t)\} = L\{x(t)\}\,G(s)$.

into a single system by addition of the transfer functions, see example 2.14. Fig. 2.4 shows the equivalent systems using block diagram notation. Fig. 2.5 shows a *feedback system* and its equivalent transfer function. For an introduction to feedback control systems, see Burghes & Graham (1980).

The idea of transfer function is closely related to the concept of *impedance* in a mechanical or electrical system. For electrical systems, taking current as input and voltage as output, the impedance function (or transfer function) for common devices is shown in Fig. 2.6, see Holbrook (1966).

The overall impedance of devices in series is obtained by *adding* their impedances (this is different from transfer functions in series, since a device 'output', voltage, is not strictly the 'input' to the next device). The derivation of transfer functions for electrical networks is demonstrated in Example 2.15.

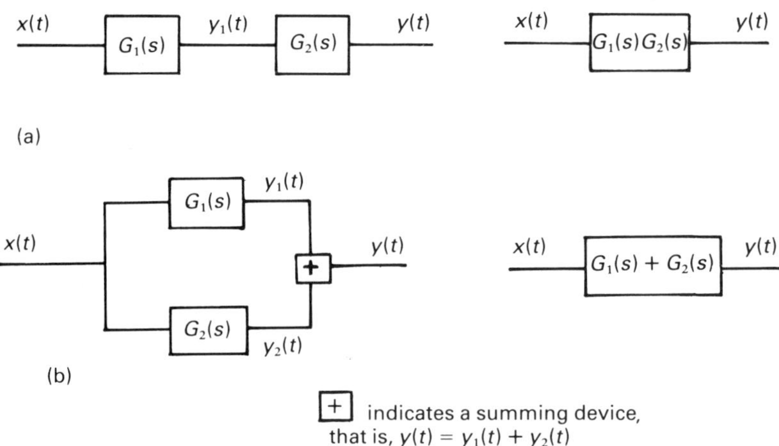

Fig. 2.4 Equivalent systems: (a) systems in series, (b) systems in parallel.

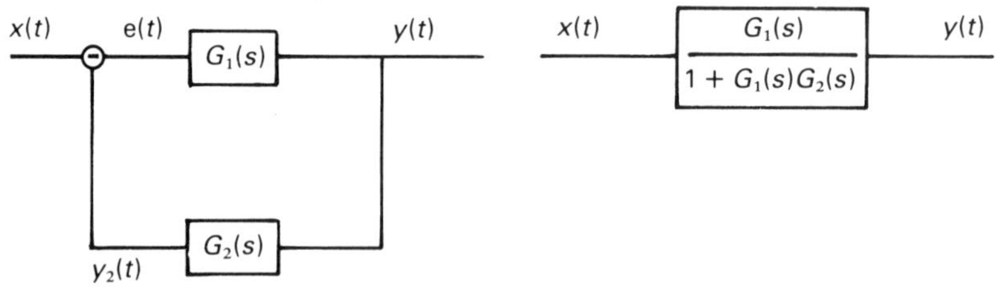

Fig. 2.5 A feedback system and its equivalent.

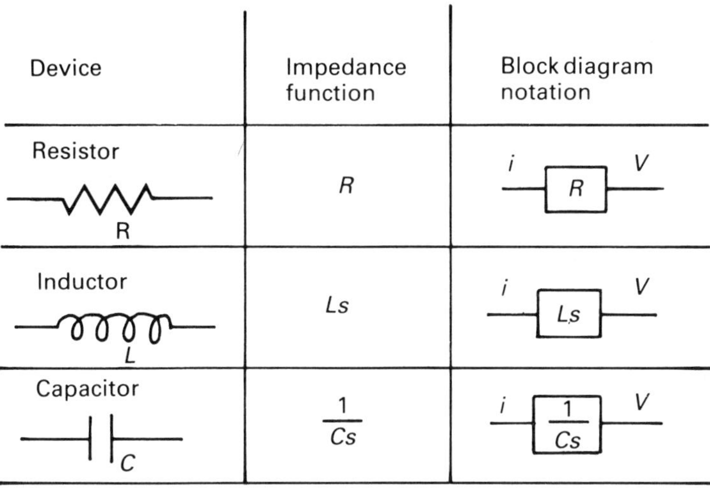

Device	Impedance function	Block diagram notation
Resistor	R	
Inductor	Ls	
Capacitor	$\dfrac{1}{Cs}$	

Fig. 2.6 Impedance functions for common devices.

Example 2.12
Express the differential equation

$$\ddot{y} + 4\dot{y} + 4y = 4\dot{x} + x$$

in block diagram form. Obtain the output $y(t)$ when the input $x(t)$ is the *unit ramp function*,

$$x(t) = \begin{cases} t & \text{when } t \geq 0 \\ 0 & \text{when } t < 0 \end{cases}.$$

Assuming that $x(0) = y(0) = \dot{y}(0) = 0$, and transforming the differential equation,

$$(s^2 + 4s + 4)L\{y\} = (4s + 1)L\{x\}.$$

Therefore the transfer function is given by $\dfrac{L\{y\}}{L\{x\}} = \dfrac{4s + 1}{s^2 + 4s + 4}$,

and in block diagram form the system is

$$x(t) \quad \boxed{\dfrac{4s + 1}{s^2 + 4s + 4}} \quad y(t)$$

For the given input function, $L\{x(t)\} = \dfrac{1}{s^2}$

and therefore $L\{y(t)\} = \dfrac{4s+1}{s^2(s^2+4s+4)}$

$$= \frac{4s+1}{s^2(s+2)^2}$$

$$= \frac{3}{4s} + \frac{1}{4s^2} - \frac{3}{4(s+2)} - \frac{7}{4(s+2)^2}.$$

Therefore $y(t) = \dfrac{3}{4} + \dfrac{t}{4} - \dfrac{3e^{-2t}}{4} - \dfrac{7te^{-2t}}{4}.$

Example 2.13

Write down the differential equation which models the system shown in Fig. 2.7.

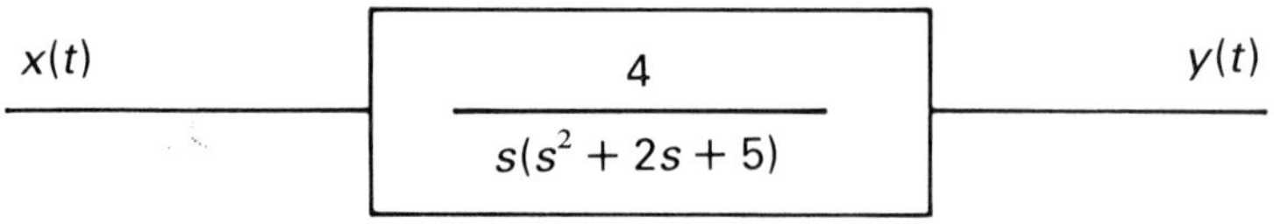

$$x(t) \quad\boxed{\dfrac{4}{s(s^2+2s+5)}}\quad y(t)$$

Fig. 2.7

The system is equivalent to the transformed equation

$$L\{y(t)\} = \frac{4}{s(s^2+2s+5)} \cdot L\{x(t)\}$$

that is, $(s^3 + 2s^2 + 5s)L\{y(t)\} = 4L\{x(t)\},$

that is, $(D^3 + 2D^2 + 5D)y(t) = 4x(t),$ where $D \equiv \dfrac{d}{dt}.$

Therefore the differential equation is

$$\dddot{y} + 2\ddot{y} + 5\dot{y} = 4x(t).$$

Example 2.14

Prove the equivalence of the systems shown in Fig. 2.4.

From Fig. 2.4(a) $L\{y_1(t)\} = L\{x(t)\} \cdot G_1(s)$

and $L\{y(t)\} = L\{y_1(t)\} \cdot G_2(s).$

Therefore $L\{y(t)\} = L\{x(t)\} \cdot G_1(s)G_2(s),$

which proves the equivalence.

From Fig. 2.4(b) $L\{y_1(t)\} = L\{x(t)\} \cdot G_1(s),$
$$L\{y_2(t)\} = L\{x(t)\} \cdot G_2(s)$$
and $L\{y(t)\} = L\{y_1(t)\} + L\{y_2(t)\}.$

Therefore $L\{y(t)\} = L\{x(t)\} \cdot G_1(s) + L\{x(t)\} \cdot G_2(s)$
$$= L\{x(t)\} \cdot (G_1(s) + G_2(s))$$
which proves the equivalence.

Example 2.15

Obtain the transfer function $\dfrac{L\{E_o(t)\}}{L\{E_1(t)\}}$ for each of the electrical networks shown in Fig. 2.8.

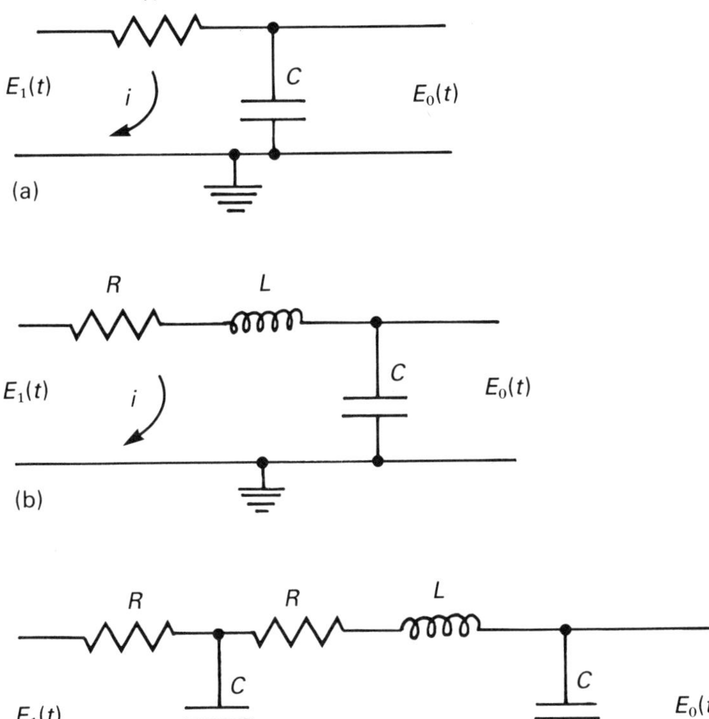

Fig. 2.8 $E_1(t)$, $E_o(t)$ are respectively input and output voltages with reference to earth.

(a) Let i be the current through the capacitor. Adding the impedance functions for the resistor and capacitor we have

$$L\{E_1(t)\} = L\{i\}\left(R + \frac{1}{Cs}\right).$$

Also $L\{E_o(t)\} = L\{i\}\,\dfrac{1}{Cs}.$

Therefore $\dfrac{L\{E_o(t)\}}{L\{E_1(t)\}} = \dfrac{1}{RCs+1}.$

(b) Adding the impedance functions for the resistor, inductor, and capacitor, we have

$$L\{E_1(t)\} = L\{i\}\left(R + Ls + \frac{1}{Cs}\right),$$

also $L\{E_o(t)\} = L\{i\}\,\dfrac{1}{Cs}.$

Therefore $\dfrac{L\{E_o(t)\}}{L\{E_1(t)\}} = \dfrac{1}{LCs^2 + RCs + 1}.$

(c) The network in Fig. 2.8(c) consists of the two previous networks in series. In block diagram notation we have

$$E_1(t) \longrightarrow \boxed{\dfrac{1}{RCs+1}} \longrightarrow \boxed{\dfrac{1}{LCs^2 + RCs + 1}} \longrightarrow E_o(t).$$

Therefore $\dfrac{L\{E_o(t)\}}{L\{E_1(t)\}} = \dfrac{1}{(RCs+1)(LCs^2 + RCs + 1)}.$

Example 2.16

Given that the input voltage in the network of Fig. 2.8(a) is given by

$$E_1(t) = \begin{cases} V \text{ when } t \geq 0,\ V \text{ constant} \\ 0 \text{ when } t < 0 \end{cases}$$

find $E_o(t)$. What is the current i at time t?

From Example 2.15(a)

$$L\{E_o(t)\} = L\{E_1(t)\}\left(\frac{1}{RCs+1}\right)$$

$$= \frac{V}{s(RCs+1)}$$

$$= \frac{V}{s} - \frac{V}{s+(RC)^{-1}}.$$

Therefore $E_o(t) = V - Ve^{-t/RC}$.

We note that $L\{E_o(t)\} = L\{i\}\dfrac{1}{Cs}$,

that is $L\{i\} = Cs\,L\{E_o(t)\}$.

This is equivalent to $i(t) = C\dfrac{\mathrm{d}E_o}{\mathrm{d}t}$.

Therefore $i(t) = \dfrac{V}{R}e^{-t/RC}$.

Problems

(1) Express the following differential equations in block diagram form, and in each case obtain the output $y(t)$ when the input $x(t)$ is a unit step function.
   (i) $\ddot{y} + 8y = \dot{x} + 16x$.

   (ii) $\ddot{y} + 6\dot{y} + 8y = \displaystyle\int_0^t x(u)\,\mathrm{d}u$.

(2) Write down the differential equations described by the diagrams shown in Fig. 2.9.

(3) Fig. 2.10(a) shows a capacitor and a resistor connected in parallel. Assuming that the current $i = i_1 + i_2$ (Kirchoff's law) show that the device has an impedance function, $\dfrac{R}{RCs+1}$. Obtain the transfer function $\dfrac{L\{E_o(t)\}}{L\{E_1(t)\}}$ for the network shown in Fig. 2.10(b).

(4) Find the output $E_o(t)$ of the network shown in Fig. 2.10(b) when the input $E_1(t)$ is a unit step function and $R_1 = R_2 = R$.

(5) A constant voltage V is applied at time $t = 0$ across a network consisting of a resistor R, an inductor L, and a capacitor C connected in series. Find an expression for the current at time t, given that $R^2C - 4L = 0$.

(6) A simple mass-spring system is shown in Fig. 2.11(a). Movement of the free

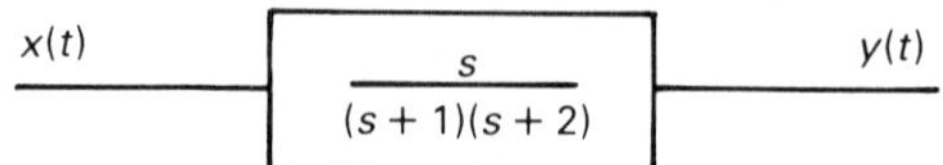

(a)

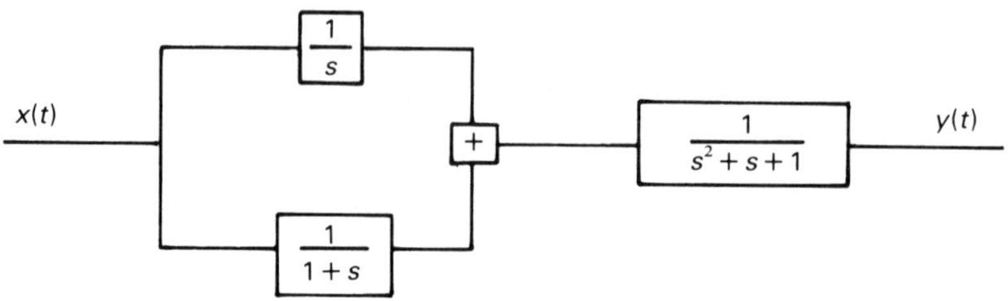

(b)

Fig. 2.9

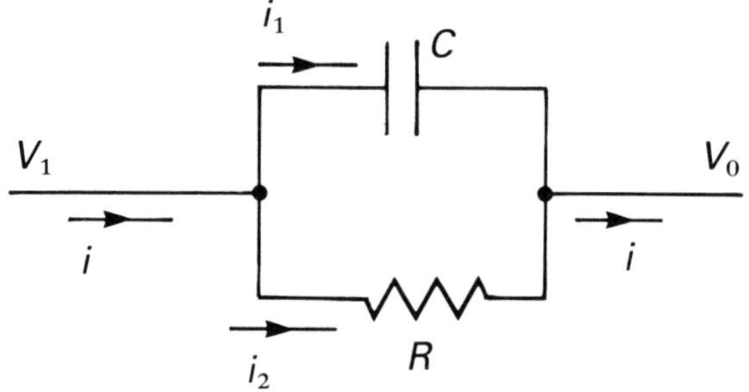

(a)

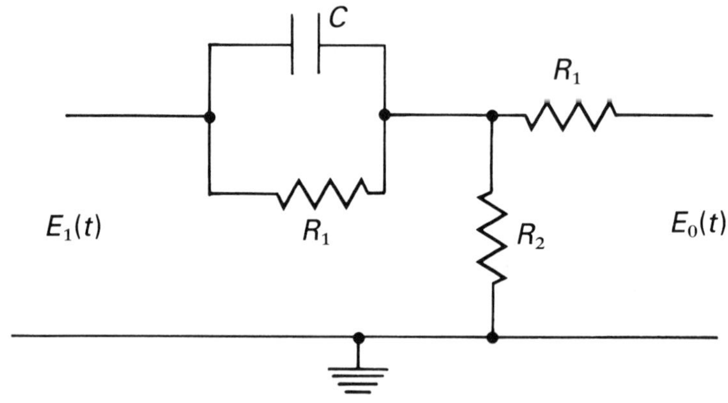

(b)

Fig. 2.10

end A is considered as the input, and the resulting movement of the mass is the output. Neglecting friction the equation of motion of the mass is

$$m\ddot{y} = -k(y-x).$$

Express this equation in block diagram form.

Obtain the transfer function $\dfrac{L\{y\}}{L\{x\}}$ of the system shown in Fig. 2.11(b) and deduce that the double spring system is equivalent to a single spring system of stiffness $(k_1^{-1} + k_2^{-1})^{-1}$. Find the output of the double spring system when x is the unit step function and $y(0) = \dot{y}(0) = 0$.

(7) Prove the equivalence of the systems shown in Fig. 2.5. Find the output of the feedback system when

$$G_1(s) = \frac{1}{s+3}, \quad G_2(s) = \frac{s+4}{s},$$

and the input is a unit step function.

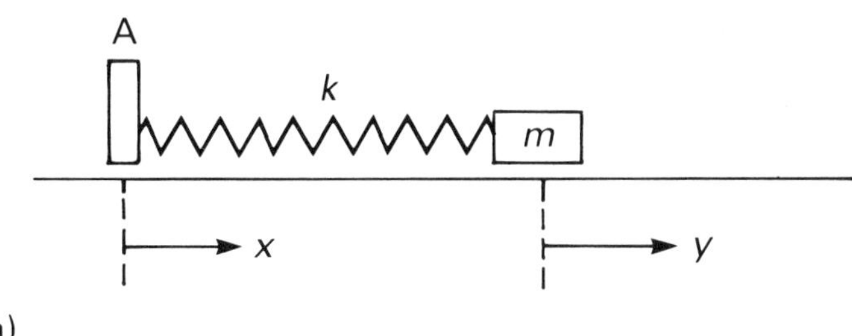

(a)

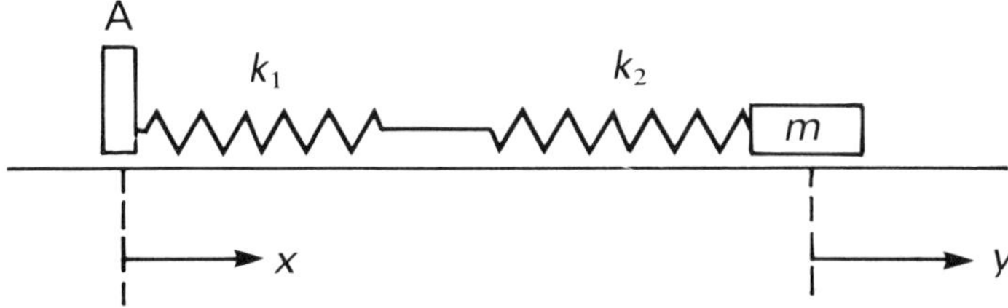

(b)

Fig. 2.11 k, k_1, k_2 are the spring stiffnesses.

Answers to problems

(1) (i)

$$x(t) \longrightarrow \boxed{\dfrac{s+16}{s^3+8}} \longrightarrow y(t).$$

When $x(t) = u(t)$

$$y(t) = 2 - \frac{7e^{-2t}}{12} + \frac{e^{t}}{12}(\sqrt{3}\sin\sqrt{3}t - 17\cos\sqrt{3}t).$$

(ii)

$$x(t) \longrightarrow \boxed{\dfrac{1}{s(s^2+6s+8)}} \longrightarrow y(t).$$

When $x(t) = u(t)$, $y(t) = \dfrac{t}{8} - \dfrac{3}{32} + \dfrac{e^{-2t}}{8} - \dfrac{e^{-4t}}{32}.$

(2) (a) $\ddot{y} + 3\dot{y} + 2y = \dot{x}$
  (b) $\dddot{y} + 2\ddot{y} + 2\ddot{y} + \dot{y} = 2\dot{x} + x.$

(3)
$$\frac{1 + R_1 Cs}{1 + \left(\dfrac{R_1 R_2}{R_1 + R_2}\right)Cs}.$$

(4) $E_0(t) = 1 + e^{-2t/RC}.$

(5) $i(t) = \dfrac{Vte^{-Rt/2L}}{L}.$

(6)

$$x \longrightarrow \boxed{\dfrac{k}{ms^2 + k}} \longrightarrow y.$$

$$y = 1 - \cos\omega t, \text{ where } \omega^2 = \frac{k_1 k_2}{m(k_1 + k_2)}.$$

(7) $y = te^{-2t}.$

2.6 STEADY-STATE AND TRANSIENT SOLUTIONS

The output from a system with transfer function $G(s)$ consists of two parts. One part is input independent and contains terms similar to those appearing in the function $g(t) = L^{-1}\{G(s)\}$. The other part is directly related to the input function. In the language of differential equations these parts are referred to respectively as 'complementary function' and 'particular integral'. Control engineers and others using systems notation use the alternative labels 'transient response' and 'steady state'.

The **transient response** of a system is that part of the output which is independent of the input. The **steady state** is the *additional* part of the output which arises owing to the presense of the input. The distinction is clearer when dealing with *stable* systems and when the input function does not tend to zero as $t \to \infty$. (We define a **stable system** as one whose impulse response function $g(t) = L^{-1}\{G(s)\}$ tends to zero as $t \to \infty$).

For this situation the transient response (as its name suggests) decays to zero as $t \to \infty$ leaving only the steady-state output. For example, consider the system with transfer function

$$G(s) = \frac{2}{(s+1)(s+2)};$$

it is a stable system since $g(t) = 2e^{-t} - 2e^{-2t} \to 0$ as $t \to \infty$. If the system input $x(t)$ is a unit step function then

$$L\{x(t)\} = \frac{1}{s}, \ L\{y(t)\} = \frac{2}{s(s+1)(s+2)}$$

and the system output is given by $y(t) = 1 - 2e^{-t} + e^{-2t}$. Here the transient response is $y_{TR}(t) = -2e^{-t} + e^{-2t}$ and the steady state is just $y_{SS}(t) = 1$.

Fig. 2.12 shows the components of the output for this example.

In Example 2.12 of section 2.5 we considered $G(s) = \dfrac{4s+1}{s^2+4s+4}$ with $x(t)$ as the unit ramp function. Here

$$y_{TR}(t) = -\frac{3}{4}e^{-2t} - \frac{7te^{-2t}}{4} \ \text{ and } \ y_{SS}(t) = \frac{3}{4} + \frac{t}{4}.$$

We have defined the stability of a system in terms of its impulse response function, but it is valuable to investigate what conditions on the system transfer function $G(s)$ are necessary for system stability. We define a **pole** of $G(s)$ as a value of s which makes $G(s)$ infinite.

Assuming that $G(s)$ can be expressed as the quotient of polynomials then a *pole* of $G(s)$ at $s = \sigma$ is equivalent to a *factor* $(s - \sigma)$ of the denominator polynomial. This in turn give a term $\dfrac{A}{(s - \sigma)}$ in the partial fraction expansion of $G(s)$ and hence a term $Ae^{\sigma t}$ in the function $g(t)$. Clearly σ must be negative for stability. If $G(s)$ has a pair of complex conjugate poles at $\sigma \pm j\omega$ then the partial fraction expansion includes the form $\dfrac{B(s - \sigma) + C\omega}{(s - \sigma)^2 + \omega^2}$ and $g(t)$ contains a term $e^{\sigma t}(B\cos \omega t + C\sin \omega t)$. Again $\sigma < 0$ is required for system stability. In general a system will be stable if *all* the poles of the system transfer function $G(s)$ lie in the left-hand half of the complex s-plane.

For systems with simple poles (that is, non-repeated) on the imaginary axis the function $g(t)$ will contain terms such as A, $B\cos \omega t$, $C\sin \omega t$ and provided

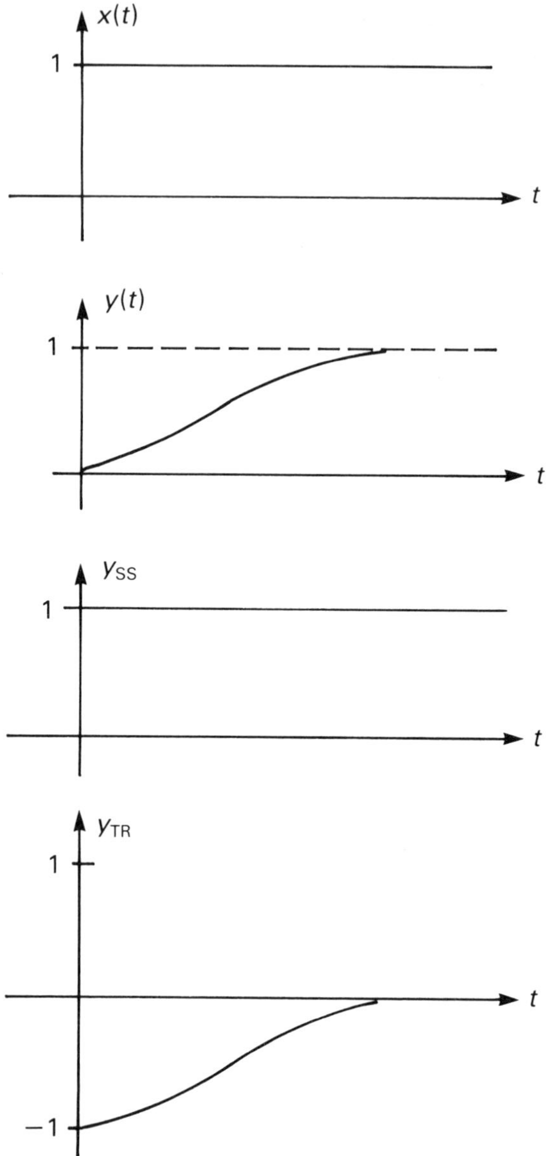

Fig. 2.12 Transient response and steady-state for the system $\dfrac{2}{(s+1)(s+2)}$.

that all other poles lie in the left-hand half-plane the system is termed **marginally stable**. In all other cases $g(t) \to \pm \infty$ as $t \to \infty$, and the system is **unstable**. The **stable region** of the complex s plane is shown in Fig. 2.13.

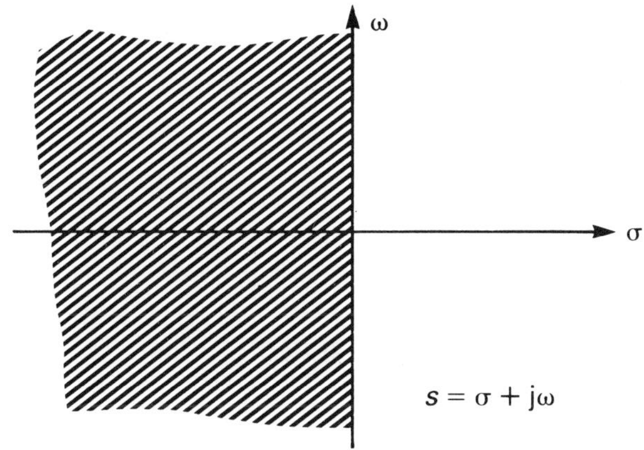

Fig. 2.13 Stable region of the s-plane.

Given that a *stable* system $G(s)$ has an input $x(t) = Ku(t)$, where K is a constant and $u(t)$ is the unit step function, then the steady-state output is simply $KG(0)$. This follows from $L\{y(t)\} = \dfrac{KG(s)}{s}$, which has the term $\dfrac{KG(0)}{s}$ in its partial fraction expansion. This result is a special case of Theorem 2.5.

Theorem 2.5 (final value theorem)

$$\lim_{t \to \infty} \{f(t)\} = \lim_{s \to 0} \{sF(s)\}, \text{ where } F(s) = L\{f(t)\},$$

assuming the limits exist.

Proof
From Theorem 2.4,

$$L\{f'(t)\} = \int_0^\infty f'(t)\,e^{-st}\,dt = sF(s) - f(0).$$

Letting $s \to 0$ and assuming the limits exist,

$$\int_0^\infty f'(t)\,dt = \lim_{s \to 0}\{sF(s) - f(0)\}.$$

Therefore $f(\infty) - f(0) = \lim_{s \to 0}\{sF(s)\} - f(0),$

that is, $\lim_{t \to \infty}\{f(t)\} = \lim_{s \to 0}\{sF(s)\}.$

Theorem 2.6 (initial value theorem)

$$\lim_{t \to 0^+} \{f(t)\} = \lim_{s \to \infty} \{sF(s)\},$$

assuming that the limits exist.

Proof

As in Theorem 2.5,

$$\int_0^\infty f'(t)\,e^{-st}\,dt = sF(s) - f(0).$$

Now let $s \to \infty$ to obtain, assuming the limits exist,

$$0 = \lim_{s \to \infty} \{sF(s) - f(0)\},$$

that is, $\lim_{t \to 0^+} \{f(t)\} = \lim_{s \to \infty} \{sF(s)\}.$

Note that we write $t \to 0^+$ since in practice we deal with functions, such as system outputs, which are zero for $t < 0$, that is, $f(0^-) = 0$.

The initial values and steady-state (when it exists) of a system output and its derivatives may be obtained using Theorems 2.5 and 2.6 without the necessity of completing the system solution, see Example 2.19. (In Chapter 3 we will consider a simple method for obtaining the steady-state output of a system when the input is an oscillatory function such as $a \sin \omega t$ or $a \cos \omega t$.)

Example 2.17

Find the output of the system with transfer function

$$G(s) = \frac{2s + 5}{s^2 + 2s + 5} \qquad \text{when the input function is}$$

(i) the unit step function,
(ii) the unit ramp function

In each case separate the steady state and the transient response.

(i) For the unit step function the input is

$$x(t) = 1 \text{ for } t \geqslant 0.$$

Therefore $L\{x(t)\} = \dfrac{1}{s}$ and

$$L\{y(t)\} = \frac{2s+5}{s(s^2+2s+5)}$$

$$= \frac{1}{s} - \frac{s}{s^2+2s+5}$$

$$= \frac{1}{s} - \frac{s+1}{(s+1)^2+2^2} + \frac{\frac{1}{2}(2)}{(s+1)^2+2^2}.$$

Therefore $y(t) = 1 - e^{-t}\cos 2t + \frac{1}{2}e^{-t}\sin 2t,$

that is, $y_{SS} = 1$, $y_{TR} = -e^{-t}(\cos 2t - \frac{1}{2}\sin 2t)$.

(ii) For the unit ramp function the input is

$$x(t) = t \text{ for } t \geqslant 0.$$

Therefore $L\{x(t)\} = \dfrac{1}{s^2}$ and

$$L\{y(t)\} = \frac{2s+5}{s^2(s^2+2s+5)}$$

$$= \frac{1}{s^2} - \frac{1}{s^2+2s+5}$$

$$= \frac{1}{s^2} - \frac{\frac{1}{2}(2)}{(s+1)^2+2^2}.$$

Therefore $y(t) = t - \dfrac{1}{2}e^{-t}\sin 2t,$

that is, $y_{SS} = t$, $y_{TR} = -\frac{1}{2}e^{-t}\sin 2t$.

Example 2.18

For the feedback system shown in Fig. 2.14, K is a *positive* constant. Show that the system is stable when $K < 2$ and unstable when $K > 2$.

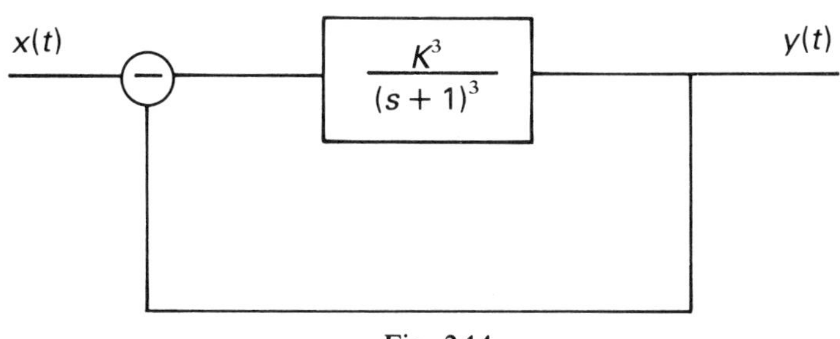

Fig. 2.14

Using the equivalent transfer function shown in Fig. 2.5,

$$\frac{L\{y(t)\}}{L\{x(t)\}} = \frac{\dfrac{K^3}{(1+s)^3}}{1 + \dfrac{K^3}{(1+s)^3}},$$

that is, the system transfer function is

$$\frac{K^3}{(1+s)^3 + K^3} = \frac{K^3}{(1+s+K)((1+s)^2 - K(1+s) + K^2)}$$

$$= \frac{K^3}{(s+K+1)(s^2 + (2-K)s + K^2 - K + 1)}$$

which has poles at $s = -(K+1)$ and

$$s = \frac{-(2-K) \pm \sqrt{(2-K)^2 - 4(K^2 - K + 1)}}{2} = \frac{-(2-K) \pm j\sqrt{3}K}{2}.$$

Therefore when $K > 2$ there are a pair of complex conjugate poles in the right-hand half of the complex s-plane, that is, the system is unstable. For $K < 2$ all the poles lie in the left-hand half-plane and the system is stable.

Example 2.19
Use Theorems 2.5 and 2.6 to calculate the following properties of the output $y(t)$ for the system with transfer function

$$G(s) = \frac{2s + 5}{s^2 + 2s + 5}.$$

(i) $y_{ss}(t)$, $y(0^+)$, $\dot{y}(0^+)$ when the input is a unit step function.
(ii) $\dot{y}_{ss}(t)$, $\ddot{y}(0^+)$ when the input is a unit ramp function.

(i) $L\{y(t)\} = \dfrac{2s + 5}{s(s^2 + 2s + 5)}$ for the step input.

$$y_{ss}(t) = \lim_{t \to \infty} y(t)$$

$$= \lim_{s \to 0} \left\{ \frac{s(2s + 5)}{s(s^2 + 2s + 5)} \right\}$$

$$= 1.$$

$$y(0^+) = \lim_{s \to \infty} \left\{ \frac{s(2s + 5)}{s(s^2 + 2s + 5)} \right\}$$

$$= 0.$$

$$L\{\dot{y}(t)\} = sL\{y(t)\}, \text{ since } y(0) = 0,$$

$$= \frac{2s+5}{s^2+2s+5}$$

$$\dot{y}(0^+) = \lim_{s \to \infty} \left\{ \frac{s(2s+5)}{s^2+2s+5} \right\}.$$

$$= 2.$$

Note the discontinuity in $\dot{y}(t)$ at $t = 0$.

(ii) $L\{y(t)\} = \dfrac{2s+5}{s^2(s^2+2s+5)}$ for the ramp input.

Therefore $L\{\dot{y}(t)\} = \dfrac{s(2s+5)}{s^2(s^2+2s+5)}$,

and $\dot{y}_{ss}(t) = \lim\limits_{t \to \infty} \dot{y}(t)$

$$= \lim_{s \to 0} \left\{ \frac{s^2(2s+5)}{s^2(s^2+2s+5)} \right\}$$

$$= 1.$$

$$L\{\ddot{y}(t)\} = s^2 L\{y(t)\}, \text{ since } y(0) = \dot{y}(0) = 0,$$

$$= \frac{2s+5}{s^2+2s+5}$$

$$\ddot{y}(0^+) = \lim_{s \to \infty} \left\{ \frac{s(2s+5)}{s^2+2s+5} \right\}.$$

$$= 2.$$

(All the answers in this example can be checked from the results of Example 2.17.)

Problems

(1) Obtain the steady-state output and transient response for the following systems with transfer function $G(s)$ and input $x(t)$.

(i) $G(s) = \dfrac{2}{(1+s)(1+2s)}$, $x(t)$ is a unit step function.

(ii) $G(s) = \dfrac{s+1}{s^2+0.5s+1}$, $x(t)$ is a unit ramp function.

(iii) $G(s) = \dfrac{1}{s^2+2s+1}$, $x(t) = 2\sin t$.

(2) Classify the stability of the following systems.

(i) $G(s) = \dfrac{1}{s^3 + 1}$.

(ii) $G(s) = \dfrac{1}{s^2 + 1}$.

(iii) $G(s) = \dfrac{1}{(s^2 + 1)^2}$.

(iv) $G(s) = \dfrac{1 - s}{s^2 + s + 1}$.

(v) $G(s) = \dfrac{3s}{s^3 + 4s^2 + 5s + 2}$.

(vi) $G(s) = \dfrac{s + 3}{4s^3 + 2s^2 + s + 3}$.

(3) The feedback system shown in Fig. 2.15 is clearly unstable when $K_2 = 0$. Is it possible to choose a value of K_2 such that the system is stable?

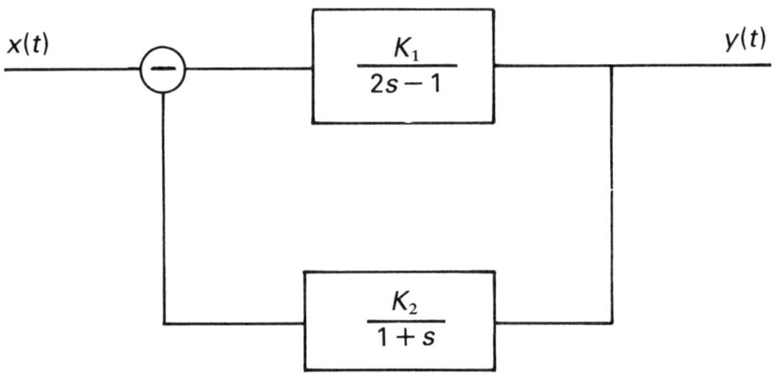

Fig. 2.15

(4) Calculate the following without considering the complete solution.

(i) $\dot{y}(0^+)$, $\ddot{y}(0^+)$, $\dot{y}_{ss}$ for system $G(s) = \dfrac{5(s + 1)}{s^2 + 0.1s + 2}$

with input $x(t) = t$, $t \geqslant 0$.

(ii) $y(0^+)$, y_{ss} for system $G(s) = \dfrac{5(s + 1)(s + 2)}{s^2 + 0.1s + 2}$

with input $x(t) = 3(1 - e^{-3t})$, $t \geqslant 0$.

(5) An important variable in a feedback control system is the error, $e(t) = x(t) - y_2(t)$, which is the difference between input and feedback, see Fig. 2.5. Show that $e(t)$ is given by

$$L\{e(t)\} = \frac{L\{x(t)\}}{1 + G_1(s)G_2(s)}$$

and write down an expression for the steady state error, e_{SS}.
   Given that $G_2(s) = 1$, show that
(i) when $x(t)$ is a step input $e_{SS} = 0$ only when $G_1(s)$ has a pole at $s = 0$,
(ii) when $x(t)$ is a ramp input $e_{SS} = 0$ only when $G_1(s)$ has a multiple pole at $s = 0$.

Show that, for the system in Fig. 2.14, $e_{SS} = \dfrac{1}{1 + K^3}$

when $x(t)$ is a unit step input.
(6) A common type of control known as 'proportional plus integral' is shown in Fig. 2.16, $K_p > 0$, $K_1 > 0$. When $x(t)$ is the unit ramp function and $H(s) = \dfrac{1}{s(s+3)}$, confirm that $e_{SS} = 0$. Show that the system can never be stable when $K_p = 0$.

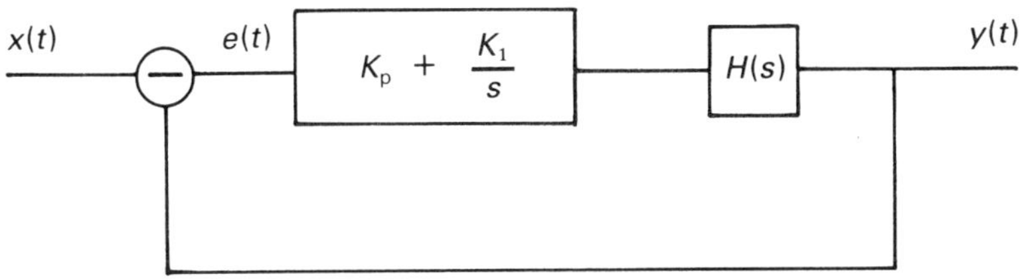

Fig. 2.16 Proportional plus integral control.

Answers to problems
(1) (i) $y_{SS} = 2$, $y_{TR} = 2e^{-t} - 4e^{-0.5t}$.

   (ii) $y_{SS} = t + 0.5$, $y_{TR} = -e^{-0.25t}\left(\dfrac{1}{2}\cos kt + \dfrac{9}{8k}\sin kt\right)$,

   where $k = \sqrt{\dfrac{15}{16}}$.

   (iii) $y_{SS} = -\cos t$, $y_{TR} = e^{-t}(1 + t)$.

(2) (i) unstable, (ii) marginally stable, (iii) unstable,
 (iv) stable, (v) stable, (vi) unstable.

(3) $K_1 K_2 > 1$ for stability.

(4) (i) $\dot{y}(0^+) = 0$, $\ddot{y}(0^+) = 5$, $\dot{y}_{SS} = \frac{5}{2}$.
 (ii) $y(0^+) = 0$, $y_{SS} = 15$.

$$(5)\quad e_{SS} = \lim_{s \to 0} \left\{ \frac{sL\{x(t)\}}{1 + G_1(s)G_2(s)} \right\}.$$

CHAPTER 3

Steady-state oscillations

3.1 INTRODUCTION

In the previous chapter we saw that the steady-state output of a stable system in response to a step input may be obtained by multiplying the level of the input by $G(0)$ where $G(s)$ is the system transfer function. A similar method is available to obtain the steady-state output when the input is a sine (or cosine) wave, and this is developed to define a graphical description of a system which can be used to write down its steady-state output when the input frequency is given. The method is particularly useful when dealing with a number of systems in series, and may be applied for more general periodic inputs by making use of the Fourier series expansion.

3.2 THE FREQUENCY RESPONSE FUNCTION

When the input to a *linear* system is given by $x(t) = a \sin \omega t$, then it is clear that the output will contain terms of frequency ω. Further, if the system is stable then the steady-state output will consist of terms in $\sin \omega t$ and $\cos \omega t$ only. For example, consider the system with transfer function

$$G(s) = \frac{1}{1+s} \text{ and input } x(t) = 2 \sin 3t.$$

The output is given by

$$L\{y(t)\} = \frac{6}{(s^2 + 3^2)(1+s)}$$

$$= \frac{0.6}{s+1} - \left(\frac{0.6s - 0.6}{s^2 + 3^2}\right).$$

Therefore $y(t) = 0.6\,e^{-t} - 0.6 \cos 3t + 0.2 \sin 3t$, and

$$y_{ss} = -0.6 \cos 3t + 0.2 \sin 3t.$$

For the general case $x(t) = a \sin \omega t$, we note that $a \sin \omega t$ is the imaginary part of $ae^{j\omega t}$, that is, we may use $ae^{j\omega t}$ as input and then take the imaginary part of the resulting output. Thus for a system with transfer function $G(s)$ the transformed output $L\{y(t)\}$ is given by the imaginary part of

$$L\{ae^{j\omega t}\}G(s) = \frac{aG(s)}{s - j\omega}$$

$$= \frac{A}{s - j\omega} + \frac{B}{s - p_1} + \frac{C}{s - p_2} + \ldots$$

where $s = p_1, p_2, \ldots$ are the poles of $G(s)$. Calculations of the constants in the partial fraction expansion gives $A = aG(j\omega)$, and the system output is given by

$$y(t) = \text{Imaginary Part} \{aG(j\omega)e^{j\omega t} + B\exp(p_1 t) + C\exp(p_2 t) + \ldots\}$$

Assuming the system is stable, the poles $p_1, p_2, \ldots$ all lie in the left-hand half of the complex s-plane, and the steady-state output is

$$y_{ss}(t) = \text{Imaginary Part} \{aG(j\omega)e^{j\omega t}\}. \tag{3.1}$$

Now, for a given value of input frequency ω, $G(j\omega)$ is a complex number which may be expressed in polar form

$$G(j\omega) = Re^{j\phi} \tag{3.2}$$

where R and ϕ depend on ω.

Substituting equation (3.2) into (3.1),

$$y_{ss}(t) = \text{Imaginary Part} \{aRe^{j\phi}e^{j\omega t}\}$$

$$= \text{Imaginary Part} \{aRe^{j(\omega t + \phi)}\}.$$

Therefore the steady-state output of the system $G(s)$ for input $x(t) = a\sin \omega t$ is given simply by

$$y_{ss}(t) = aR\sin(\omega t + \phi) \tag{3.3}$$

where R and ϕ are the modulus and argument of the complex number obtained by putting $s = j\omega$ in the system transfer function. Once again we see the correspondence between the Laplace parameter s and the frequency variable ω.

The following points should be noted.

(i)   The *frequency* of the output is the same as the input frequency.

(ii)   The *amplitude* of the output is obtained by scaling the input amplitude by a factor R. R is called the **gain** of the system $G(s)$, denoted by $|G(j\omega)|$.

(iii) There is a *phase difference* of amount ϕ between the output and input, which is equivalent to a time shift of the waveform of amount ϕ/ω. ϕ is called the **phase** of the system $G(s)$, denoted by $\lfloor G(j\omega)$.

(iv) The steady-state output of the system $G(s)$ for input $x(t) = a\cos \omega t$ is obtained by taking the *real* part in equation (3.1) to give $y_{ss}(t) = aR\cos(\omega t + \phi)$.

The function $G(j\omega)$ is called the **frequency response function** of the system. It is in fact the *Fourier transform* of the system's impulse response function. Using the frequency response function we are able to write down the steady-state oscillatory output whenever the input is a sine (or cosine) wave of given frequency.

Example 3.1

Find the steady-state output of the system $G(s) = \dfrac{1}{1+0.1s}$ for the following inputs:

(i) $x(t) = 2\sin 10t$ (ii) $x(t) = 2\sin 50t$

$$G(j\omega) = \frac{1}{1+0.1\,\omega j} = R\mathrm{e}^{j\phi}$$

where $R = \dfrac{1}{\sqrt{(1+0.01\,\omega^2)}}$, $\phi = -\tan^{-1}(0.1\,\omega)$.

(i) When $\omega = 10$, $R = \dfrac{1}{\sqrt{2}}$, $\phi = -\dfrac{\pi}{4}$

$$\text{Therefore } y_{\mathrm{ss}}(t) = 2\left(\frac{1}{\sqrt{2}}\right)\sin\left(10t - \frac{\pi}{4}\right)$$

$$= \sqrt{2}\sin\left(10t - \frac{\pi}{4}\right),$$

that is, the amplitude has been scaled down by a factor $\dfrac{1}{\sqrt{2}}$ and there is a

phase lag amount $\dfrac{\pi}{4}$. If we write the output as $y_{\mathrm{ss}}(t) = \sqrt{2}\sin 10\left(t - \dfrac{\pi}{40}\right)$ we

see that the output wave 'lags' the input wave by an amount $\dfrac{\pi}{40}$ on the time

axis; that is, the $\dfrac{\pi}{4}$ phase lag is equivalent to a time axis lag of $\dfrac{T}{8}$ where $T = \dfrac{\pi}{5}$ is

the period of the input function.

(ii) When $\omega = 50$, $R = \dfrac{1}{\sqrt{26}} \simeq 0.196$,

$$\phi = -\tan^{-1}5 \simeq -1.373.$$

Therefore $y_{\mathrm{ss}}(t) \simeq 0.392\sin(50t - 1.373)$.

For this frequency the amplitude has been scaled down by a factor 0.196, and the phase lag is equivalent to a time lag of 0.0275.

Example 3.2

Obtain the steady-state output of the system $G(s) = \dfrac{1}{s^2 + s + 1}$ with input $x(t) = a \cos \omega t$, $t \geqslant 0$. For what value of ω is maximum amplification obtained? Write down the output for this case.

$$G(j\omega) = \frac{1}{1 - \omega^2 + j\omega} = R e^{j\phi},$$

where $R = \dfrac{1}{\sqrt{(1 - \omega^2)^2 + \omega^2}}$,

and $\phi = \begin{cases} -\tan^{-1}\left(\dfrac{\omega}{1 - \omega^2}\right) & \text{when } \omega < 1 \\[2ex] -\pi + \tan^{-1}\left(\dfrac{\omega}{\omega^2 - 1}\right) & \text{when } \omega > 1 \\[2ex] -\dfrac{\pi}{2} & \text{when } \omega = 1, \end{cases}$

and $y_{ss}(t) = aR \cos(\omega t + \phi)$.

The amplification factor is R, and this is maximized when $(1 - \omega^2)^2 + \omega^2$ is minimised. That is,

when $\dfrac{d}{d\omega}((1 - \omega^2)^2 + \omega^2) = 0$.

Therefore $-4\omega + 4\omega^3 + 2\omega = 0$,

which is satisfied by $\omega = 0$ and $\omega = \dfrac{1}{\sqrt{2}}$.

For $\omega = 0$ the input is a *step function*, and the amplification factor is just $G(0) = 1$.

For $\omega = \dfrac{1}{\sqrt{2}}$ the amplification factor is $R = \dfrac{1}{\sqrt{0.75}} \simeq 1.155$, and $\phi = -\tan^{-1}(\sqrt{2}) \simeq -0.955$.

Thus the maximum amplification is obtained when $\omega = \dfrac{1}{\sqrt{2}}$ and the corresponding steady state output is $y_{ss}(t) \simeq 1.155\, a \cos\left(\dfrac{t}{\sqrt{2}} - 0.955\right)$.

Problems

(1) Obtain expressions for $|G(j\omega)|$ and $\underline{/G(j\omega)}$ for the system

$$G(s) = \frac{s+1}{s^2+4s+4}.$$

Write down the output of the system for each of the following inputs,
(i) $3 \sin 2t$,
(ii) $4 \cos 10t$,

(iii) $\sin\left(t + \dfrac{\pi}{3}\right)$.

(2) Describe briefly the effect of the system

$$G(s) = \frac{1+s}{1+2s}$$

on the input $a \sin \omega t$ when
(i) $\omega \ll 1$,
(ii) $\omega \gg 10$.
At what frequency is the phase lag maximised?

(3) Write down the solution for large t of the differential equation

$$\ddot{y} + 2\dot{y} + y = 4 \sin 2t.$$

(4) The system $G(s) = \dfrac{\omega_n^2}{s^2 + 2\delta_n \omega_n s + \omega_n^2},\ 0 < \delta_n < \dfrac{1}{\sqrt{2}},$

has input $x(t) = a \sin \omega t$, (ω_n and δ_n are constants known respectively as natural frequency and damping factor). For what value of ω is the output amplitude maximized? For this frequency find the value of δ_n to give an amplification factor of 10.

(5) For the electrical network shown in Fig. 2.10(b) find the steady-state output voltage E_o when $E_1 = 10 \sin 3t$ and $R_1 C = R_2 C = 0.2$. What is the steady output when the input frequency is
(i) very low,
(ii) very high?

Answers to problems

(1) $|G(j\omega)| = \sqrt{\dfrac{1+\omega^2}{(4-\omega^2)^2 + 16\omega^2}}$

$$
\lfloor G(j\omega) = \begin{cases} \tan^{-1}\omega - \tan^{-1}\left(\dfrac{4\omega}{4-\omega^2}\right) & \text{when } \omega < 2 \\[3ex] \tan^{-1}\omega - \pi + \tan^{-1}\left(\dfrac{4\omega}{\omega^2-4}\right) & \text{when } \omega > 2 \\[3ex] \tan^{-1}2 - \dfrac{\pi}{2} & \text{when } \omega = 2. \end{cases}
$$

(i) $\quad -\dfrac{3\sqrt{5}}{8}\cos\left(2t+\gamma\right)$ where $\gamma = \tan^{-1}2$.

(ii) $\quad -\sqrt{\dfrac{101}{676}}\cos\left(10t+\beta\right)$ where $\beta = \tan^{-1}10 + \tan^{-1}\left(\dfrac{5}{12}\right)$.

(iii) $\quad \dfrac{\sqrt{2}}{5}\sin\left(t+\alpha\right)$ where $\alpha = \dfrac{7\pi}{12} - \tan^{-1}\left(\dfrac{4}{3}\right)$.

(2) (i) output $\simeq$ input for small ω.

(ii) output $\simeq \frac{1}{2}$(input) for large ω.

$$\omega = 1\sqrt{2} \text{ to maximize phase lag.}$$

(3) $y_{ss} = -0.8\sin\left(2t+\alpha\right)$ where $\alpha = \tan^{-1}\left(\dfrac{4}{3}\right)$

or $y_{ss} = -0.48\sin 2t - 0.64\cos 2t$.

(4) $\omega = \omega_n\sqrt{1-2\delta_n^2}$, $\delta_n \simeq 0.05$.

(5) $E_{oss} \simeq 11.17\sin\left(3t+0.249\right)$.

(i) $E_{oss} \simeq 10\sin\omega t$ when ω small.

(ii) $E_{oss} \simeq 20\sin\omega t$ when ω large.

3.3 BODE DIAGRAMS

The graphs of $|G(j\omega)|$ and $\lfloor G(j\omega)$ against ω are called the **Bode diagrams** for the system $G(s)$. Together the two graphs are an alternative representation of the system since the functions $|G(j\omega)|$ and $\lfloor G(j\omega)$ completely define the system transfer function $G(j\omega)$. Given the Bode diagrams for a system and the input $x(t) = a\sin\omega t$, the steady-state output $y_{ss}(t) = aR\sin\left(\omega t + \phi\right)$ can be written down by simply reading off from the graphs $R = |G(j\omega)|$ and $\phi = \lfloor G(j\omega)$. The Bode diagrams for the first order system $G(s) = \dfrac{1}{(1+sT)}$ are shown in Fig. 3.1, using the *non-dimensional form* ωT on the horizontal axis. It is convenient to

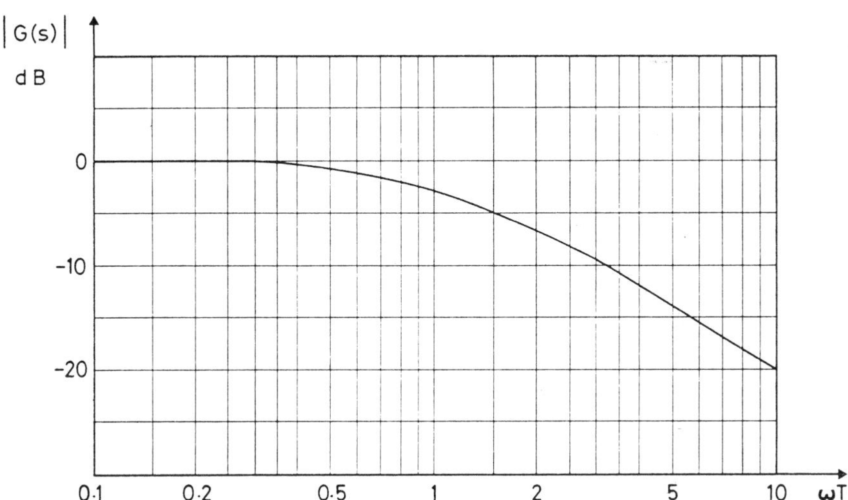

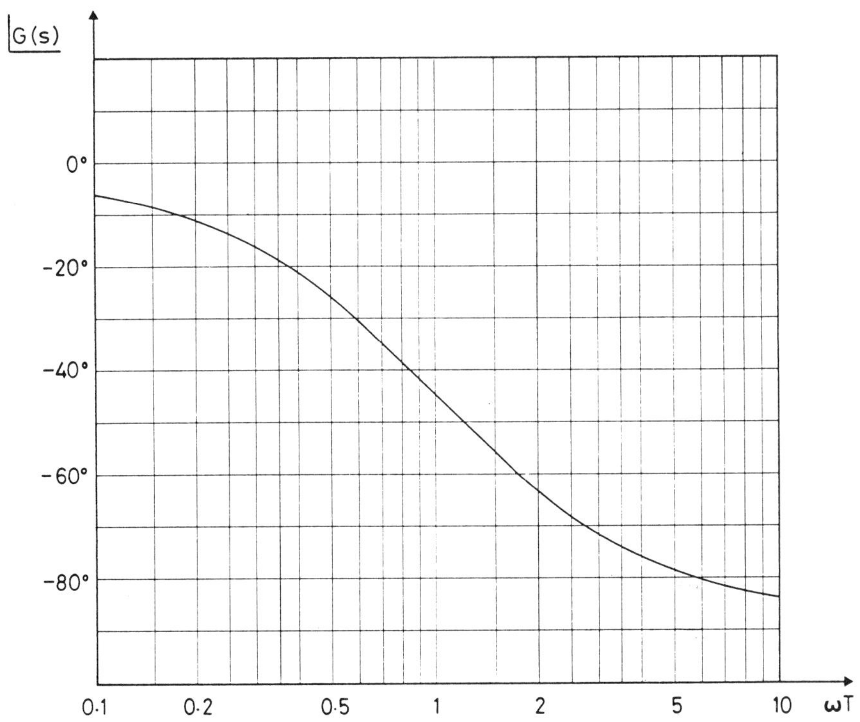

Fig. 3.1 Bode diagram for $G(s) = \dfrac{1}{1+sT}$.

plot ωT on a logarithmic scale and to present $|G(j\omega)|$ in the dB logarithmic form†, defined by

$$R \, \text{dB} = 20 \log_{10} R.$$

The reason for this log-log presentation is partly because the graph of $|G(j\omega)|$ is a straight line for large ω, but more importantly because the *gains* of two systems in series may be *added* when in dB form. This follows because for systems $G_1(j\omega) = R_1 \exp(j\phi_1)$ and $G_2(j\omega) = R_2 \exp(j\phi_2)$ in series the equivalent system $G(j\omega) = R \exp(j\phi)$ is given by $G(j\omega) = G_1(j\omega) G_2(j\omega) = R_1 R_2 \exp(j(\phi_1 + \phi_2))$. That is, $\phi = \phi_1 + \phi_2$ and $R = R_1 R_2$, but in dB form $R \, \text{dB} = R_1 \, \text{dB} + R_2 \, \text{dB}$. Therefore both the gain R *and* the phase ϕ of the equivalent system $G(j\omega)$ can be obtained by addition.

The Bode diagrams for the second order system $G(s) = \dfrac{\omega_n^2}{s^2 + 2\delta_n \omega_n s + \omega_n^2}$

are shown in Fig. 3.2 for a range of values of δ_n; the nondimensional form ω/ω_n is used on the horizontal axis. ω_n and δ_n are called the **natural frequency** and **damping factor** of the system.

Usually the transfer function of a system can be expressed as the quotient of polynomials, and since polynomials can be factorized into linear and quadratic factors, Figs 3.1 and 3.2 may be used to obtain the steady-state oscillatory output of the system, see Example 3.4. In certain circumstances the transfer function of a particular system may be unknown, but, assuming the system is linear, it is still possible to obtain the Bode diagrams by experiment using a transfer function analyser. This equipment generates a sine wave input and measures the amplification and phase shift of the output for a range of input frequencies. The resulting Bode diagrams may give some information about the system's transfer function. For example, the slope of the $|G(j\omega)|$ plot for large ω is related to the difference in degree of the numerator and denominator polynomials of $G(s)$. However, if the system is just one component of a large system, then the steady-state oscillations of the latter may be determined by consideration of Bode diagrams; it is not necessary to know the transfer functions of every component.

Example 3.3
Use Fig. 3.1 to obtain the steady-state output of the system

$$G(s) = \frac{1+s}{1+2s} \qquad \text{when the input is } x(t) = 5 \sin 2t.$$

---

† dB denotes *decibels*, units used by systems engineers to measure amplitude of waves. Note that $R = 2$ is equivalent to $R \, \text{dB} \simeq 6$ and $R = \tfrac{1}{2}$ is equivalent to $R \, \text{dB} \simeq -6$.

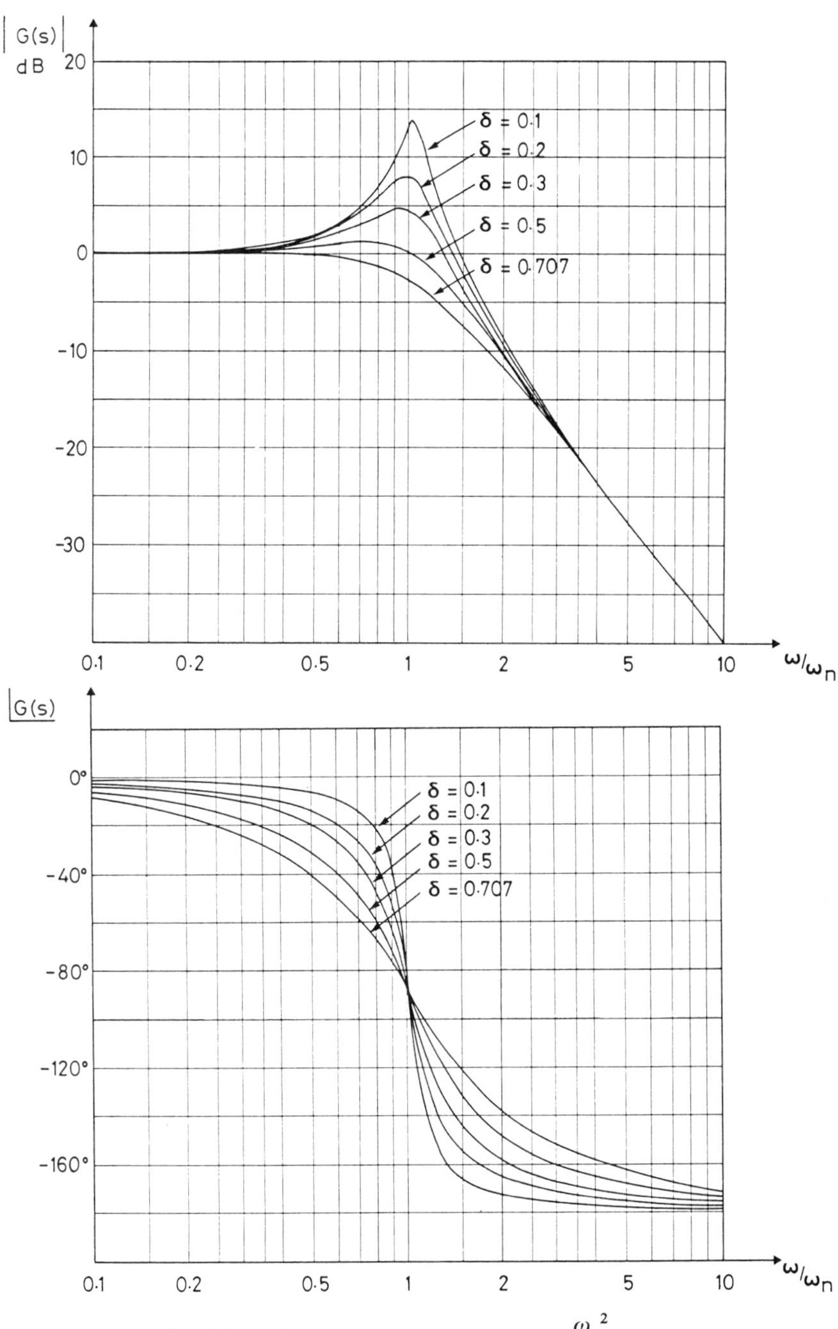

Fig. 3.2 Bode diagram for $G(s) = \dfrac{\omega_n^2}{s^2 + 2\delta_n \omega_n s + \omega_n^2}$.

The system $G(s)$ can be considered as two systems, $G_1(s) = \dfrac{1}{1+2s}$ and $G_2(s) = 1 + s$ in series. The gain and phase of $G_1(s)$ can be read directly from Fig. 3.1 for $\omega T = 2 \times 2 = 4$. The gain and phase of $G_2(s)$ are obtained by applying Fig. 3.1 to the function $\dfrac{1}{1+s}$, $\omega T = 2 \times 1 = 2$, and *changing the sign* of both $R\,\mathrm{dB}$ and ϕ, since $\dfrac{1}{G_2(s)} = \dfrac{1}{R_2 \exp(j\phi_2)} = \dfrac{1}{R_2}\exp(-j\phi_2)$ and

$$\frac{1}{R_2}\,\mathrm{dB} = -R_2\,\mathrm{dB}.$$

Therefore $\quad\quad |G_1(j2)| = -12.3\,\mathrm{dB}, \quad \underline{|G_1(j2)} = -76°$

and $\quad\quad\quad |G_2(j2)| = 7\,\mathrm{dB}, \quad\quad \underline{|G_2(j2)} = 63.5°,$

Therefore $\quad\quad |G(j2)| = -12.3 + 7 = -5.3\,\mathrm{dB}.$

and $\quad\quad\quad \underline{|G(j2)} = -76° + 63.5° = -12.5°.$

It follows that the gain factor $R = 0.543$ and the phase shift $\phi = -0.218$ (radians), that is, the steady-state output is

$$y_{\mathrm{ss}}(t) = 2.71 \sin(2t - 0.218) \text{ approximately.}$$

Example 3.4
Use Figs 3.1 and 3.2 to obtain the steady-state output of the system

$$G(s) = \frac{50(s + 20)}{s^3 + 14s^2 + 140s + 1000}$$

when the input is $x(t) = \cos 20t$, $t \geqslant 0$.

Factorizing the denominator of $G(s)$ gives

$$G(s) = \frac{50(s + 20)}{(s + 10)(s^2 + 4s + 100)} = \frac{100(1 + 0.05s)}{(1 + 0.1s)(s^2 + 4s + 100)}.$$

We can consider $G(s)$ as equivalent to the following three systems in series.

$G_1(s) = 1 + 0.05s,$ $\quad\quad$ that is, $T = 0.05,$

$G_2(s) = \dfrac{1}{1 + 0.1s}$ $\quad\quad$ that is, $T = 0.1,$

$G_3(s) = \dfrac{100}{s^2 + 4s + 100}$ $\quad\quad$ that is, $\omega_\mathrm{n} = 10,\ \delta_\mathrm{n} = 0.2.$

For the given input function the frequency $\omega = 20$, and from Fig. 3.1

$$|G_1(j20)| = 3 \text{ dB}, \qquad \underline{G_1(j20)} = 45°$$

$$|G_2(j20)| = -7 \text{ dB}, \qquad \underline{G_2(j20)} = -63.5°$$

From Fig. 3.2

$$|G_3(j20)| = -9.8 \text{ dB}, \qquad \underline{G_3(j20)} = -165.1°.$$

Therefore since $G(s) = G_1(s) \cdot G_2(s) \cdot G_3(s)$.

$$|G(j20)| = 3 - 7 - 9.8 = -13.8 \text{ dB (that is, } R = 0.20)$$

$$\underline{G(j20)} = 45° - 63.5° - 165.1° = -183.6° \text{ (that is, } \phi = -3.20)$$

and the steady-state output is

$$y_{SS}(t) = 0.2 \cos(20t - 3.2) \text{ approximately.}$$

Example 3.5

A system $G(s)$ consists of two subsystems $G_1(s)$ and $G_2(s)$ connected in series.
$G_1(s) = \dfrac{1}{1 + 0.02s}$, but $G_2(s)$ is unknown. However, Bode diagrams for the
second subsystem were obtained by experiment, and the following table
contains particular readings from these graphs.

Frequency ω rad/sec	$|G_2(j\omega)|$	$\underline{G_2(j\omega)}$
0	12 dB	0°
10	13 dB	-34°
150	-23 dB	-172°

Find the output of the system $G(s)$ for the input
$x(t) = 2 + \sin 10t + 0.5 \sin 150t$.

Using Fig. 3.1 we can construct the corresponding table for $G_1(s)$ and hence
$G(s)$ since $G(s) = G_1(s) \cdot G_2(s)$.

ω	$|G_1(j\omega)|$	$\underline{G_1(j\omega)}$
0	0 dB	0°
10	-0.2 dB	-11.3°
150	-10 dB	-71.6°

ω	$|G(j\omega)|$	$\underline{G(j\omega)}$
0	12 dB	0°
10	12.8 dB	-45.3°
150	-33 dB	-243.6°

The corresponding amplification factors and phase shifts for the system $G(s)$ are given below.

ω	R	ϕ (radians)
0	4	0
10	4.4	-0.79
150	0.02	-4.25

Therefore the steady-state output of the system $G(s)$ is

$$y_{ss}(t) = 8 + 4.4 \sin(10t - 0.79) + 0.01 \sin(150t - 4.25).$$

(The linearity property allows us to add the outputs from each component of the input).

Problems
(1) Use Figs 3.1 and 3.2 to confirm the results of Problems 1, 3, and 5 in section 3.2.
(2) Apply Figs 3.1 and 3.2 to obtain the steady-state output of the system

$$G(s) = \frac{s^2 + 16s + 100}{(s+4)(s^2 + 4s + 16)}$$

when the input function is $x(t) = 1 - 2\cos 6t$.
(3) Solve the differential equation

$$\ddot{y} + 3\dot{y} + 2y = 3\dot{x} + 2x$$

where $x = \cos 2t$ and $y(0) = \dot{y}(0) = 0$. Use Fig. 3.1 to confirm the steady-state solution.
(4) Control engineers often approximate to $|G(j\omega)|$ for

$$G(s) = \frac{1}{1 + sT} \qquad \text{as follows,}$$

$$|G(j\omega)| = \begin{cases} 0 \text{ dB for } \omega T < 1 \\ -20 \log_{10}(\omega T) \text{ dB for } \omega T > 1 \end{cases}.$$

Show that the maximum error in the gain factor R caused by this approximation is just above 40%. At what frequency does the maximum error occur?
(5) The guidance system of an air-to-air missile measures $\dot{\phi}$, the rate of change of the line of sight to target. The output from the guidance system is a voltage V which is input to the missile control system to produce a lateral

acceleration $\ddot{y}$. An approximation to the guidance system is

$$\frac{L\{V\}}{L\{\phi\}} = \frac{2}{1+0.1\,s} \text{ volts/rad. sec.}^{-1},$$

and the missile control system is

$$\frac{L\{\ddot{y}\}}{L\{V\}} = G(s),$$

where $G(j\omega)$ has been determined by experiment in the form of Bode diagrams. In particular $G(0) = 300\,\mathrm{ms}^{-2}/\mathrm{volt}$,

$$\left|\frac{G(j5)}{G(0)}\right| = 2\ \mathrm{dB}, \underline{\lfloor G(j5)} = -40°. \text{ Given that the target manoeuvres in such}$$

a way that $\dot{\phi} = 0.1\,(1-\cos 5t)$, find the steady-state lateral acceleration of the missile.

(6) The pressure of liquid in a pipeline is varying sinusoidally (that is, as a sine wave) between 1500 psi and 3000 psi at a frequency of 10 Hz (that is, $\omega = 20\pi$). A pressure gauge connected to the pipeline may be considered as

a first ordèr system $G(s) = \dfrac{1}{1+0.2s}$.

Calculate the upper and lower readings on the gauge. Repeat the calculation for another type of gauge which may be treated as a second order system with natural frequency 500 Hz and damping factor 0.8.

Answers to problems
(2) $y_{\mathrm{ss}}(t) = 1.56 - 1.02\cos(6t - 2.3)$ approximately.
(3) $y = 2e^{-2t} - 2.8e^{-t} + 0.8\cos 2t + 0.6\sin 2t$.
(5) $\ddot{y} = 60 - 67\cos(5t - 1.16)$ approximately.
(6) 2310 psi and 2190 psi, approximately.
      2999.9 psi and 1500.1 psi, approximately (that is, use second gauge).

3.4 THE ANALOG FILTER

Consider the electronic network shown in Fig. 2.8(a) with $R = 10^5$ Ohms, $C = 2 \times 10^{-6}$ Farads, and input voltage $E_1(t) = 2\sin \pi t + 0.5\sin 100\pi t$. Using the result of Example 2.15(a), the network is described by the transfer function

$$G(s) = \frac{1}{1 + RCs} = \frac{1}{1 + 0.2s}.$$

Now $\left|G(j\omega)\right| = \dfrac{1}{\sqrt{1+0.04\omega^2}}$ and $\underline{\lfloor G(j\omega)} = -\tan^{-1}(0.2\omega).$

Therefore $|G(j\pi)| = 0.847$, $\underline{/G(j\pi)} = -0.561$,
and $|G(j100\pi)| = 0.016$, $\underline{/G(j100\pi)} = -1.555$,
and the steady-state network output is

$$E_o(t) = 1.694 \sin(\pi t - 0.561) + 0.008 \sin(100\pi t - 1.555), \text{ approx.}$$

Therefore the high-frequency (50 Hz) component has been virtually removed by the action of the network. We say that the network has 'filtered out' the high-frequency content of the input function. There are many everyday applications where such filtering is necessary. For example consider the measurement of pressure in a pipeline as in Problem 6 of section 3.3. Suppose that variation of pressure includes a low-amplitude high-frequency component in addition to the main variation of 10 Hz. The frequency response characteristics of the measuring instrument may tend to amplify this high-frequency component to such an extent that it is impossible to read the instrument's display. It is therefore necessary to include an electronic network before the measuring instrument to remove the unwanted high-frequency signal. A device chosen to modify the frequency characteristics of a *continuous* signal in some specified way is known as an **analog filter**. In particular the first order system $G(s) = \dfrac{1}{1+Ts}$ passes at least 90% of input components of frequency less that $0.077\,T^{-1}$ Hz, and passes less that 10% of components of frequency greater than $1.584\,T^{-1}$ Hz. The first order system is an approximation to what is called a *low-pass filter*. The gain $\sim$ frequency diagram for an *ideal* low-pass filter is shown in Fig. 3.3(a). Other common filter requirements are shown in Figs 3.3(b), (c), (d). Transfer functions $G(s)$ which approximate to these ideal characteristics are well documented in the literature; an introduction to filter synthesis is given by Stanley (1975). In this section only particular examples are considered. In Chapter 8 the concept of an analog filter is used for the design of *digital filters*, that is, filters which process *discrete* input functions.

Example 3.6
Consider the second order system with transfer function

$$G(s) = \frac{10^5}{s^2 + 10^2 s + 10^6}.$$

Show that the maximum value of $|G(j\omega)|$ is approximately 1, and find the corresponding value of ω. Obtain the frequency range for which $|G(j\omega)| < 0.2$. Comment on the application of this system as an analog filter.

$$G(j\omega) = \frac{10^5}{10^6 - \omega^2 + j10^2\omega}$$

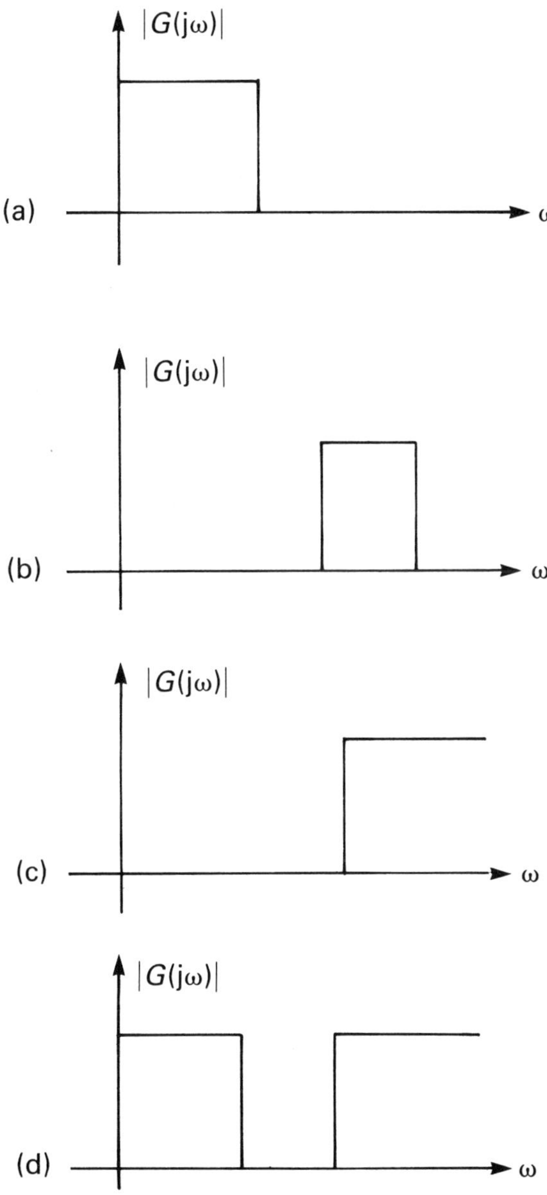

Fig. 3.3 Ideal filter gain characteristics (a) low-pass, (b) band-pass, (c) high-pass, (d) band rejection.

and $|G(j\omega)| = \dfrac{10^5}{\sqrt{((10^6 - \omega^2)^2 + 10^4\omega^2)}}$

which will be maximized when $(10^6 - \omega^2)^2 + 10^4\omega^2$ is minimized. Differentiating gives $-4\omega(10^6 - \omega^2) + 2 \times 10^4\omega = 0$ for maximum $|G(j\omega)|$. Therefore $\omega^2 = 10^6 - 0.5 \times 10^4$ for which value

$$|G(j\omega)| = \frac{10^5}{\sqrt{(10^{10} - 0.25 \times 10^8)}}.$$

Therefore the maximum value of $|G(j\omega)|$ is approximately 1, and the corresponding frequency is $\omega = 997.5$ radians per second.

For $|G(j\omega)| < 0.2$ we have $\dfrac{10^{10}}{(10^6 - \omega^2)^2 + 10^4\omega^2} < 0.04$

which can be written

$\omega^4 - 1.99 \times 10^6\omega^2 + 0.75 \times 10^{12} > 0$, that is,

$\omega^2 < 505\,000$ or $\omega^2 > 1\,485\,000$ approximately.

Therefore $|G(j\omega)| < 0.2$ for all frequencies *outside* the range $711 < \omega < 1219$ radians/second. The system $G(s)$ behaves as a *band-pass filter* operating in a narrow band around $\omega = 1000$. If the input consists of a summation of equal amplitude sine waves over a wide frequency range then the output will be dominated by terms of frequency near $\omega = 1000$. The filter is said to be *tuned* to the frequency $\omega = 1000$.

An electronic network corresponding to the above filter can be designed using a circuit similar to the one shown in Fig. 2.8(b).

Example 3.7
An approximation to the ideal low-pass filter is given by

$$|G(j\omega)|^2 = \frac{1}{1 + (\omega/\omega_c)^{2k}}$$

where k is the order of the corresponding transfer function $G(s)$, and ω_c is the cut-off frequency. Obtain the transfer function $G(s)$ for the cases $k = 1, 2, 3$.

For $k = 1$, $G(s)$ is clearly the first order transfer function

$$\frac{1}{1 + Ts}, T = \frac{1}{\omega_c}, \text{ since}$$

$$|G(j\omega)| = \frac{1}{\sqrt{1 + \omega^2 T^2}} = \frac{1}{\sqrt{1 + (\omega/\omega_c)^2}}.$$

For $k = 2$ we have $|G(j\omega)|^2 = \dfrac{1}{1 + (\omega/\omega_c)^4}$.

For any complex number z with conjugate $\bar{z}$ we have $z\bar{z} = |z|^2$.

Therefore $G(j\omega)\, G(-j\omega) = |G(j\omega)|^2 = \dfrac{1}{1 + \omega^4/\omega_c^{\,4}}$.

Recalling that $s = j\omega$ we have $G(s)\, G(-s) = \dfrac{\omega_c^{\,4}}{s^4 + \omega_c^{\,4}}$.

Assuming that $G(s)$ represents a *stable* system, then all the poles of $G(s)$ lie in the left-hand half of the complex s-plane, and consequently the poles of $G(-s)$ lie (symmetrically placed) in the right-hand half-plane. Therefore we can obtain $G(s)$ by factorising $G(s)G(-s)$ and selecting that part which corresponds to stable poles.

$$G(s)\, GR(-s) =$$
$$\frac{\omega_c^{\,4}}{(s+(1+j)\omega_c/\sqrt{2})(s+(1-j)\omega_c/\sqrt{2})(s+(-1+j)\omega_c/\sqrt{2})(s+(-1-j)\omega_c/\sqrt{2})}$$

Therefore
$$G(s) = \frac{\omega_c^{\,2}}{(s+(1+j)\omega_c/\sqrt{2})(s+(1-j)\omega_c/\sqrt{2})}$$
$$= \frac{\omega_c^{\,2}}{(s^2 + \sqrt{2}\,\omega_c s + \omega_c^{\,2})}.$$

Following the same method for $k = 3$,
$$G(s)\, G(-s) = \frac{\omega_c^{\,6}}{\omega_c^{\,6} - s^6}.$$

The poles of $G(s)\, G(-s)$ are at $s = \pm\omega_c$,
$$s = \omega_c \exp\left(\pm j\frac{\pi}{3}\right) = \omega_c\left(\frac{1}{2} \pm j\frac{\sqrt{3}}{2}\right) \quad \text{and}$$
$$s = \omega_c \exp\left(\pm j\frac{2\pi}{3}\right) = \omega_c\left(-\frac{1}{2} \pm j\frac{\sqrt{3}}{2}\right).$$

Selecting only the poles in the left-hand half-plane we have
$$G(s) = \frac{\omega_c^{\,3}}{(s+\omega_c)\left(s+\omega_c\left(\frac{1}{2}+j\frac{\sqrt{3}}{2}\right)\right)\left(s+\omega_c\left(\frac{1}{2}-j\frac{\sqrt{3}}{2}\right)\right)}$$
$$= \frac{\omega_c^{\,3}}{s^3 + 2\omega_c s^2 + 2\omega_c^{\,2} s + \omega_c^{\,3}}.$$

The $|G(\mathrm{j}\omega)|^2 \sim \omega$ graphs for $k = 1, 2, 3$ are shown in Fig. 3.4.

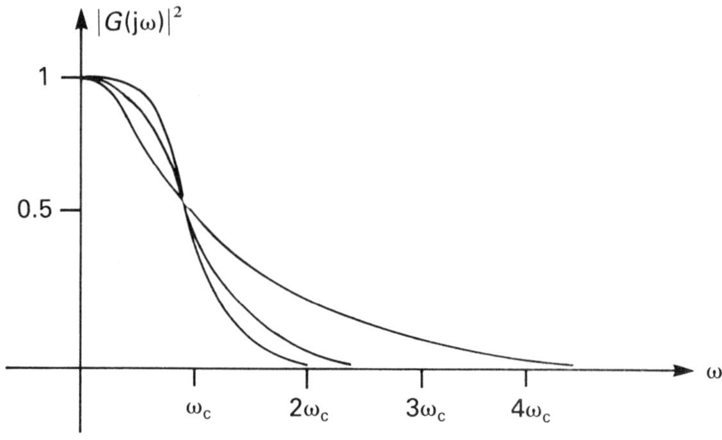

Fig. 3.4

Example 3.8

To stabilize a certain feedback control system it is necessary to increase the phase in the feedback path using the analog filter

$$G(s) = \frac{1 + 2Ts}{1 + Ts}.$$

Calculate the largest phase advance achieved by this filter and the corresponding frequency.

$$G(\mathrm{j}\omega) = \frac{1 + 2\mathrm{j}\omega T}{1 + \mathrm{j}\omega T}$$

and $\underline{/G(\mathrm{j}\omega)} = \tan^{-1} 2\omega T - \tan^{-1} \omega T$.

The maximum value for the phase is obtained from

$$\frac{\mathrm{d}}{\mathrm{d}\omega}\underline{/G(\mathrm{j}\omega)} = \frac{2T}{1 + 4\omega^2 T^2} - \frac{T}{1 + \omega^2 T^2} = 0, \quad \text{that is,}$$

$$2(1 + \omega^2 T) - (1 + 4\omega^2 T^2) = 0.$$

Therefore $\omega = \dfrac{1}{T\sqrt{2}}$ is the frequency corresponding to the maximum phase advance which is $\tan^{-1}\sqrt{2} - \tan^{-1}\dfrac{1}{\sqrt{2}} = 19.5°$.

Problems

(1) Design a first order low-pass filter $G(s)$ such
that $|G(0)| = 1$,
$\quad |G(j\omega)| > 0.5$ for $\omega < 20\pi$.
and $|G(j\omega)| < 0.5$ for $\omega > 20\pi$.

(2) The filter $G(s) = \dfrac{s}{s^2 + s + 1}$ is a crude attempt at a band pass filter. It has the
following properties
(i) $|G(j\omega)|$ is maximum when $\omega = \omega_c$,
(ii) $|G(j\omega)| > 0.5 |G(j\omega_c)|$ for $\omega_1 < \omega < \omega_2$.
Find the values of ω_c, ω_1, and ω_2.

(3) Find the transfer function $G(s)$ of the low-pass filter satisfying

$$|G(j\omega)|^2 = \frac{1}{1 + \omega^8}.$$

(4) It is necessary to design an analog filter $G(s)$ satisfying

$$|G(j\omega)|^2 = \frac{25(4 - \omega^2)^2}{(9 + \omega^2)(16 + \omega^2)}.$$

Further, the filter must be stable and *minimum phase* (that is, all poles *and
zeros* of $G(s)$ lie in the left-hand half-plane; a minimum phase filter
minimizes the delay between input and output time functions). Find $G(s)$.

(5) A phase-lag filter is required to modify the performance of a control
system. The filter transfer function is of the form $G(s) = \dfrac{K(1 + Ts)}{1 + \lambda Ts}, \lambda > 1$,

and the following conditions have to be satisfied.
(i)   The phase lag $-\underline{/G(j\omega)}$, has a maximum value of $30°$.
(ii) The maximum phase lag occurs at $\omega = 100$ radians/second.
(iii) $|G(j100)| = 1$.
Find the values of λ, T, and K.

Answers to problems

(1) $G(s) = \dfrac{1}{1 + 0.02757s}$

(2) $\omega_c = 1$, $\omega_1 = 0.457$, $\omega_2 = 2.189$.

(3) $G(s) = \dfrac{1}{s^4 + 2.613s^3 + 3.414s^2 + 2.613s + 1}$.

(4) $G(s) = \dfrac{5(s^2 + 4)}{(s + 3)(s + 4)}$.

(5) $\lambda = 3, T = 0.00577, K = \sqrt{3}$.

3.5 USE OF FOURIER SERIES

We have seen in this chapter how system transfer functions and Bode diagrams may be used to write down the steady-state output when the input is a linear combination of sine and cosine functions. The method can be extended to a more general periodic input function by using its Fourier series expansion and applying the frequency response technique term by term to the infinite series.

For example, if the input to the system

$$G(s) = \frac{1}{1+Ts} \text{ is the square wave function,}$$

$$x(t) = \begin{cases} 1 & \text{when } 0 \leqslant t < \pi \\ 0 & \text{when } \pi \leqslant t < 2\pi \end{cases}, \quad x(t+2\pi) = x(t),$$

then we write the input in the form

$$x(t) = 0.5 + \frac{2}{\pi} \sum_{r=1}^{\infty} \frac{1}{(2r-1)} \sin(2r-1)t, \qquad \text{(Example 1.1)}.$$

The steady-state system output is then

$$y_{SS}(t) = 0.5\, G(0) + \frac{2}{\pi} \sum_{r=1}^{\infty} \frac{R_{2r-1}}{(2r-1)} \sin\{(2r-1)\,t + \phi_{2r-1}\}$$

where $R_n = |G(jn)| = \dfrac{1}{\sqrt{1+n^2 T^2}}$

and $\phi_n = \lfloor G(jn) = -\tan^{-1}(nt)$.

In Chapter 4 we will study an alternative method of solution using a closed form for the Laplace transform of a periodic function. However, the above technique is clearly useful for cases where the system transfer function is unknown but a Bode diagram description is available.

Further study of the technique reveals an interesting bonus. When the system input is *non-periodic* it is possible to obtain the complete output, *steady-state plus transient response*, by considering the steady-state output to a suitably chosen periodic input.

Consider again the system

$$G(s) = \frac{1}{1+Ts} \text{ with input } x(t) = u(t), \text{ the unit step function.}$$

Suppose it is sufficient to know the system output for the time interval $0 \leqslant t \leqslant t_1$. We now take the input to be a periodic square wave $x_p(t)$ defined by

$$x_p(t) = \begin{cases} 1 & \text{when } 0 \leqslant t < L \\ 0 & \text{when } L \leqslant t < 2L \end{cases}, \quad x_p(t+2L) = x_p(t).$$

Provided that we choose the half-period L to be greater than t_1, then $x_p(t)$ and $u(t)$ are *identical* for the interval of interest $0 \leqslant t \leqslant t_1$. We will show that, provided that L/T is large enough, the *complete* solution of the system for input $u(t)$ is obtained by finding the *steady-state* solution for input $x_p(t)$.

The Fourier series expansion of the square wave function is

$$x_p(t) = 0.5 + \frac{2}{\pi} \sum_{r=1}^{\infty} \frac{1}{(2r-1)} \sin (2r-1) \frac{\pi t}{L}. \tag{3.4}$$

It is easily verified that the transient response of $G(s) = \dfrac{1}{1+Ts}$ for inputs 0.5 and $\sin nt$ are $-0.5e^{-t/T}$ and $\dfrac{n}{T}\left(\dfrac{1}{T^2}+n^2\right)^{-1} e^{-t/T}$ respectively. Therefore when the input is $x_p(t)$, the transient response is given by

$$y_{TR} = -0.5\,e^{-t/T} + \frac{2}{LT}\sum_{r=1}^{\infty} \left(\frac{1}{T^2} + (2r-1)^2 \frac{\pi^2}{L^2}\right)^{-1} e^{-t/T}$$

$$= \frac{2}{LT} e^{-t/T} \left\{ -\frac{LT}{4} + \frac{LT}{4}\left(\frac{1-e^{-L/T}}{1+e^{-L/T}}\right) \right\}$$

using the result of Problem 5 in section 1.2.

Therefore $y_{TR} = -\dfrac{e^{-L/T}\,e^{-t/T}}{(1+e^{-L/T})}$, and we note that when $L \gg T$ the transient response is negligible. For the system $G(s) = \dfrac{1}{1+Ts}$ the interval of interest, $0 \leqslant t \leqslant t_1$, is such that t_1 is of the same order as T. Therefore the condition $L \gg T$ implies that $L > t_1$ and the function $x_p(t)$ is equivalent to the unit step function. Thus for $L \gg T$ the output of $G(s) = \dfrac{1}{1+Ts}$ for a unit step function input can be obtained by writing down the *steady-state* output for input $x_p(t)$, since y_{TR} is negligble. In general, a stable system's transfer function can be expressed as the sum of the terms of the form $\dfrac{1}{1+Ts}$, and the above method may be applied provided that L is chosen to be considerably greater than the largest T. (If T is complex we use $|T| = \dfrac{1}{\omega_n}$ when deciding on the choice of L; ω_n is the natural frequency corresponding to the quadratic factor.)

Clearly in practice we do *not* use this technique when the system transfer function is *known*, but the method is useful when the system description is given in the form of Bode diagrams. In this case $R_n = |G(jn\pi/L)|$ and $\phi_n = |G(jn\pi/L)$ can be read from the Bode diagrams for $n = 1, 3, 5, 7, \ldots,$

and using Equation (3.4) the system output for a unit step is given by

$$y(t) = 0.5 R_0 + \frac{2}{\pi} \sum_{r=1}^{n} \frac{R_{2r-1}}{(2r-1)} \sin\left\{ (2r-1)\frac{\pi t}{L} + \phi_{2r-1} \right\} \qquad (3.5)$$

Example 3.9
The transfer function of a certain control system is unknown, but Bode diagrams are available, from which the following table has been obtained.

ω	0	1	3	5	7
$|G(j\omega)|$	1	1.053	1.001	0.410	0.220
$\underline{G(j\omega)}$	$0°$	$-20.6°$	$-90°$	$-137°$	$-152°$

ω	9	11	13	15	17
$|G(j\omega)|$	0.117	0.077	0.055	0.039	0.032
$\underline{G(j\omega)}$	$-159°$	$-164°$	$-166°$	$-168°$	$-170°$

Use Equation (3.5) to obtain the system output for a unit step input given that the interval of interest is $0 \leqslant t \leqslant 1$.

For convenience choose $L = \pi$. The input function is a square wave of period 2π and is identical to the unit step function for $0 \leqslant t \leqslant 1$. Also we can assume that the largest *time constant T* arising from the system is less than 1, hence L is considerably greater than T.

The calculation is set out as follows, denoting $|G(j\omega)|$ by R_ω and $\underline{G(j\omega)}$ by ϕ_ω.

ω	R_ω	ϕ_ω	$\dfrac{R_\omega}{\omega}\sin\phi_\omega$	$\dfrac{R_\omega}{\omega}\sin(0.1\omega+\phi_\omega)$	$\dfrac{R_\omega}{\omega}\sin(0.2\omega+\phi_\omega)$
1	1.053	-0.359	-0.370	-0.270	-0.167
3	1.001	-1.571	-0.334	-0.319	-0.275
5	0.410	-2.391	-0.056	-0.078	-0.081
7	0.220	-2.652	-0.015	-0.029	-0.030
9	0.117	-2.775	-0.005	-0.012	-0.011
11	0.077	-2.862	-0.002	-0.007	-0.004
13	0.055	-2.897	-0.001	-0.004	-0.001
15	0.039	-2.932	-0.001	-0.003	0
17	0.032	-2.967	0	-0.002	0.001
TOTAL			-0.784	-0.724	-0.568

Multiplying by $\dfrac{2}{\pi}$ and adding 0.5 gives

		$y(0) = 0.001$	$y(0.1) = 0.039$	$y(0.2) = 0.138$
ω	$\dfrac{R_\omega}{\omega}\sin(0.3\omega+\phi_\omega)$	$\dfrac{R_\omega}{\omega}\sin(0.4\omega+\phi_\omega)$	$\dfrac{R_\omega}{\omega}\sin(0.5\omega+\phi_\omega)$	$\dfrac{R_\omega}{\omega}\sin(0.6\omega+\phi_\omega)$
1	-0.062	0.043	0.148	0.254
3	-0.207	-0.121	-0.024	0.076
5	-0.064	-0.031	0.009	0.047
7	-0.016	0.005	0.024	0.031
9	-0.001	0.010	0.013	0.006
11	0.003	0.007	0.003	-0.004
13	0.004	0.003	-0.002	-0.004
15	0.003	0	-0.003	-0.001
17	0.002	-0.001	-0.001	0.002
TOTAL	-0.338	-0.085	0.167	0.407

Multiplying by $\dfrac{2}{\pi}$ and adding 0.5 gives

		$y(0.3) = 0.285$	$y(0.4) = 0.446$	$y(0.5) = 0.606$	$y(0.6) = 0.759$
ω	$\dfrac{R_\omega}{\omega}\sin(0.7\omega+\phi_\omega)$	$\dfrac{R_\omega}{\omega}\sin(0.8\omega+\phi_\omega)$	$\dfrac{R_\omega}{\omega}\sin(0.9\omega+\phi_\omega)$	$\dfrac{R_\omega}{\omega}\sin(\omega+\phi_\omega)$
1	0.352	0.464	0.542	0.630
3	0.168	0.246	0.302	0.330
5	0.073	0.082	0.070	0.042
7	0.024	0.006	-0.015	-0.029
9	-0.005	-0.012	-0.010	-0.001
11	-0.007	-0.002	0.005	0.007
13	0	0.004	0.002	-0.003
15	0.002	0.001	-0.002	-0.001
17	0.001	-0.002	0	0.002
TOTAL	0.608	0.787	0.894	0.977

Multiplying by $\dfrac{2}{\pi}$ and adding 0.5 gives

$$y(0.7) = 0.887 \qquad y(0.8) = 1.001 \qquad y(0.9) = 1.069 \qquad y(1.0) = 1.122$$

The graph of $y(t) \sim t$ is shown in Fig. 3.5. (The tables for $|G(j\omega)|$ and $\underline{|G(j\omega)}$ in this example correspond very closely to the function $G(s) = \dfrac{9}{s^2 + 3s + 9}$. The reader should confirm this by using Fig. 3.2 and calculating the output of this particular second order system for a unit step input, that is, $y(t) = L^{-1}\left\{\dfrac{9}{s(s^2 + 3s + 9)}\right\}$. Comparison of this function with the approximate calculations in this example shows very close agreement).

The hand calculation as set out in Example 3.9 is rather tedious, but the routine is clearly suitable for solution using a simple computer program. The author has chosen to include a large amount of numerical detail in order to demonstrate the rate at which the Fourier series converges.

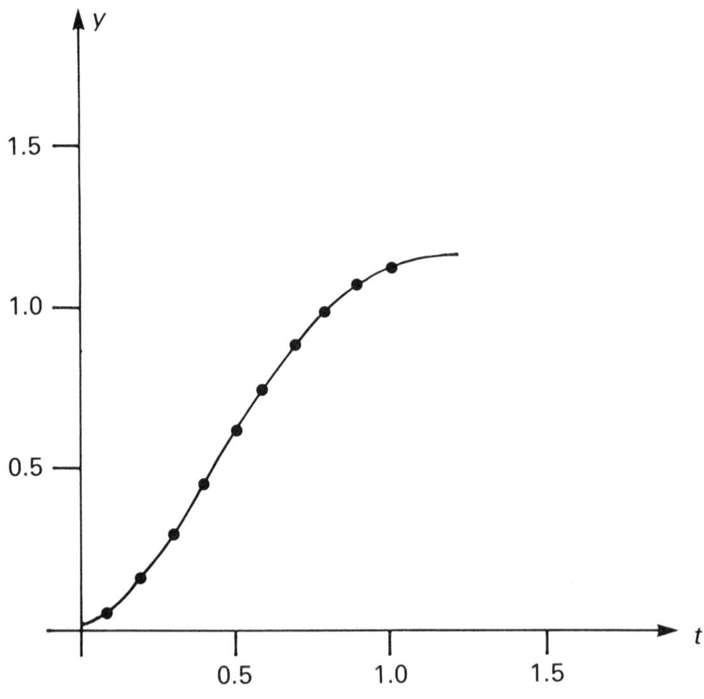

Fig. 3.5

Problems

(1) Use Equation (3.5) to obtain the output of the first order system $G(s) = \dfrac{2}{1+0.4\,s}$ for a unit step input and compare your answer with the function $L^{-1}\left\{\dfrac{2}{s(1+0.4\,s)}\right\}$.

(2) The system described in Example 3.9 is modified by filtering its input using the analog filter $\dfrac{1}{1+0.4\,s}$. Use Equation (3.5) to obtain a graph of the output of the modified system when the input is a unit step function.

3.6 RESONANCE

We have already seen the way in which a second order system can be described in terms of the parameters *natural frequency*, ω_n and *damping factor* δ_n. The system may be represented by a differential equation

$$\ddot{y} + 2\delta_n \omega_n \dot{y} + \omega_n^2 y = \omega_n^2,$$

or by a transfer function

$$G(s) = \frac{\omega_n^2}{s^2 + 2\delta_n \omega_n s + \omega_n^2}.$$

When the input is $x = a \sin \omega t$ the steady-state output is, following the notation of section 3.2,

$$y_{ss} = aR \sin (\omega t + \phi)$$

where $R = |G(j\omega)| = \dfrac{\omega_n^2}{\sqrt{[(\omega_n^2 - \omega^2)^2 + 4\delta_n^2 \omega_n^2 \omega^2]}}$

and $\tan \phi = \tan \underline{|G(j\omega)} = -\dfrac{2\delta_n \omega_n \omega}{\omega_n^2 - \omega^2}$

The result of Problem 4 in section 3.2 shows that the amplification factor R is largest when $\omega = \omega_n \sqrt{(1 - 2\delta_n^2)}$ assuming that $\delta_n < \dfrac{1}{\sqrt{2}}$. The corresponding value of R is $R_{max} = \dfrac{1}{2\delta_n \sqrt{1 - \delta_n^2}} \simeq \dfrac{1}{2\delta_n}$ when δ_n is small. The frequency $\omega_n \sqrt{(1 - 2\delta_n^2)}$ is called the **resonant frequency** of the system, that is, the input frequency which maximizes the amplitude of steady-state oscillations. For a lightly damped system (δ_n small) the output amplitude can be very large indeed and may cause certain variables to reach limiting values (such as a piston in a hydraulic system hitting the end of the cylinder). In more severe cases system destruction may occur as, for example, in the Tacoma Narrows bridge disaster in 1940, see Braun (1975). In general, for mechanical systems **resonance**, that is, oscillations at resonant frequency, should be avoided by careful design. However, in certain electrical systems the phenomenon can be used to advantage (for example in tuned circuits).

The analysis of lightly damped systems is often carried out by assuming them to have *zero* damping. This assumption greatly simplifies the analysis for high order systems, see Huntley & Johnson (1983). The undamped second order system has transfer function $G(s) = \dfrac{\omega_n^2}{s^2 + \omega_n^2}$, the resonant frequency is $\omega = \omega_n$, and it follows that $|G(j\omega_n)|$ is *infinite*. For input $x = a \sin \omega_n t$ the output is

$$y = L^{-1} \left\{ \frac{a\omega_n^3}{(s^2 + \omega_n^2)^2} \right\} = \frac{a}{2} (\sin \omega_n t - \omega_n t \cos \omega_n t),$$

and it is clear that the output amplitude becomes infinite as $t \to \infty$. Higher order systems which are *marginally stable* (that is, $G(s)$ has a *simple* pole on the imaginary axis) also exhibit this property. The denominator of $G(s)$ will have a non-repeated factor of the form $s^2 + k^2$, and for input $x = a \sin kt$ the amplitude of output oscillations will tend to infinity. For lightly damped systems the natural frequency and resonant frequency are virtually the same for each mode of oscillation. The assumption of zero damping allows the *critical frequencies* to be more readily obtained.

Example 3.10

Write down the approximate steady-state output $y_{ss}(t)$ for the system described by the differential equation

$$\ddot{y} + 16\dot{y} + 40\,000y = 120\,000 \sin 200t$$

If we take the input to be $x = 3 \sin 200t$, then the system transfer function is

$$G(s) = \frac{40\,000}{s^2 + 16s + 40\,000}$$

Therefore $\omega_n = 200$ and $\delta_n = 0.04$. Since δ_n is very small, the natural frequency and resonant frequency are virtually equal, and the amplification factor is given by $\dfrac{1}{2\delta_n} = 12.5$. The phase at the natural frequency is $-90°$ (see Fig. 3.2), and therefore the steady-state output is

$$y_{ss} = 12.5 \times 3 \sin \left(200t - \frac{\pi}{2} \right) = -37.5 \cos 200t.$$

(Alternatively we can write

$$G\,(j\,200) = \frac{40\,000}{3200j} = \frac{12.5}{j}$$

Therefore $R = 12.5$ and $\phi = -90°$).

Example 3.11
Consider the double spring system shown in Fig. 3.6. Assuming linear springs and very small damping, obtain an approximation to the transfer function $\dfrac{L\{y_1\}}{L\{x\}}$. If the point A is moved up and down at frequency ω radians/second, find approximate values for the critical frequencies for the case

$$\frac{k_1}{m_1} = 12, \qquad \frac{k_2}{m_1} = \frac{k_2}{m_2} = 8.$$

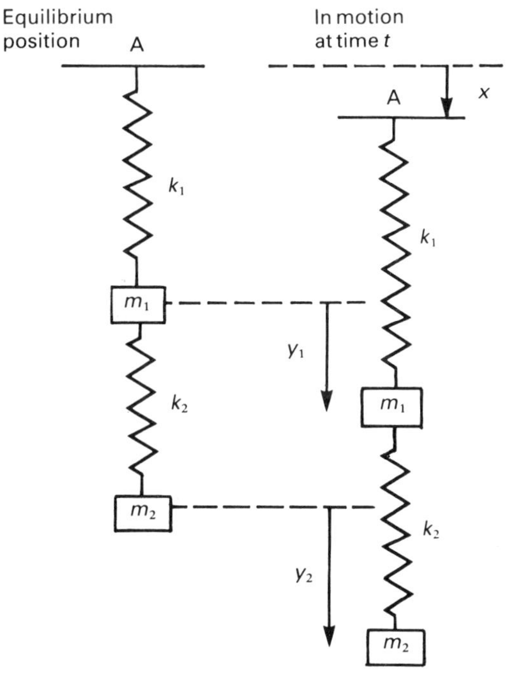

Fig. 3.6 k_1, k_2 are the spring stiffnesses; m_1, m_2 are masses.

Making the simplifying assumption of zero damping and applying Newton's second law to each mass gives

$$m_1\ddot{y}_1 = k_2(y_2 - y_1) - k_1(y_1 - x)$$
$$m_2\ddot{y}_2 = -k_2(y_2 - y_1).$$

Taking Laplace transforms (assuming zero initial conditions)

$$(m_1 s^2 + k_1 + k_2)L\{y_1\} - k_2 L\{y_2\} = k_1 L\{x\}$$
$$-k_2 L\{y_1\} + (m_2 s^2 + k_2)L\{y_2\} = 0.$$

Therefore

$$((m_1 s^2 + k_1 + k_2)(m_2 s^2 + k_2) - k_2{}^2)L\{y_1\} = k_1(m_2 s^2 + k_2)L(x\}$$

and

$$\frac{L\{y_1\}}{L\{x\}} = \frac{k_1/m_1(s^2 + k_2/m_2)}{s^4 + (k_1/m_1 + k_2/m_1 + k_2/m_2)s^2 + (k_1/m_1)(k_2/m_2)}.$$

From the given numbers, the denominator of the transfer function is

$$s^4 + 28s^2 + 96 = (s^2 + 4)(s^2 + 24).$$

Therefore the critical frequencies are $\omega = 2$ and $\omega = \sqrt{24}$. Since the damping is small, the critical frequencies for the damped system will be very near these values.

Problems
(1) Write down the approximate steady-state solution of the differential equation

$$\ddot{y} + 0.5\dot{y} + 100y = 7\sin nt,$$

where n is the resonant frequency.
(2) The position of a component in a mechanical system subject to vibration at frequency ω is given by the solution of the differential equation

$$\ddddot{x} + 9140\ddot{x} + 1.3 \times 10^7 x = 1000\sin \omega t.$$

What are the critical frequencies for the system?
(3) For the circuit shown in Fig. 3.7 the applied voltage is $E(t) = 250\sin 300t$ volts. Show that the current i amps satisfies the differential equation

$$L\frac{d^2 i}{dt^2} + R\frac{di}{dt} + \frac{i}{C} = \frac{dE}{dt}.$$

Given that $\sqrt{\dfrac{1}{LC}} = 300$, $L = 11$, and $R = 40$, will a 13 amp fuse be adequate for this circuit?

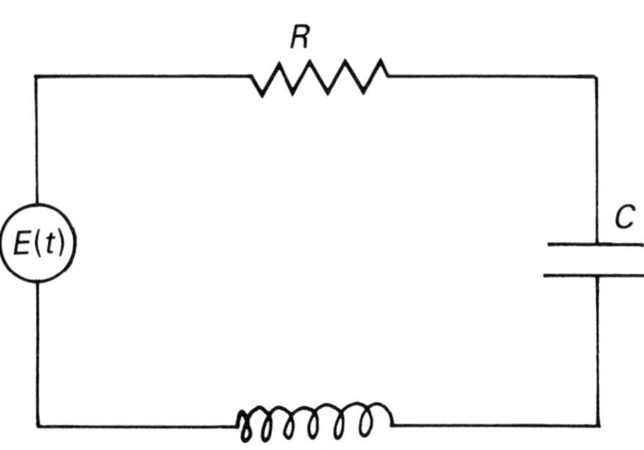

Fig. 3.7

(4) Show that the system $G(s) = \dfrac{100}{s^2 + 2s + 100}$ with input $x = 5\sin \omega t$ at resonant frequency, has approximate steady-state output $y = -25\cos \omega t$. If the input is filtered using a first order filter $\dfrac{1}{1 + sT}$ find the value of T which reduces the amplification factor to unity.

(5) A simplified model of a car travelling along a bumpy road is shown in Fig. 3.8. Suppose that the mass M of the car and spring stiffness k are such that $M/k = 0.01\,s^2$. Neglecting damping, show that the displacement $y(t)$ of the

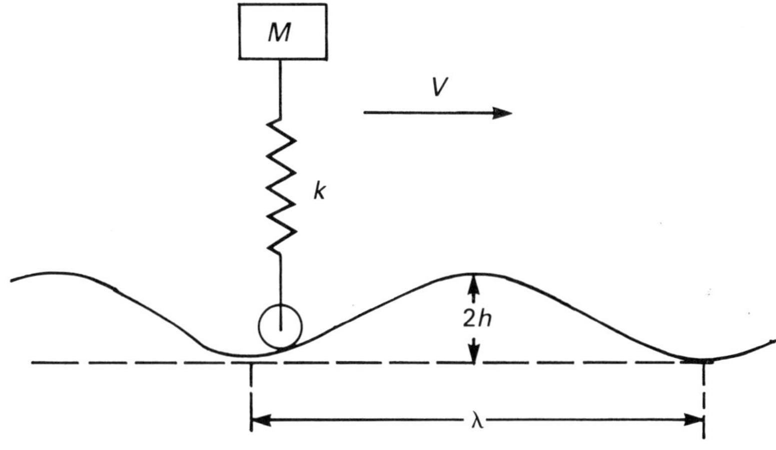

Fig. 3.8

car from its equilibrium position satisfies the differential equation

$$0.01\ddot{y} + y = h \cos\left(\frac{2\pi Vt}{\lambda}\right)$$

where V is the constant speed of the car, and the road is assumed to have cosine wave profile of amplitude h and wavelength λ. Deduce that for a road of wavelength 15 m, the worst vibrations occur when $V \simeq 24$ m/s.

Answers to problems
(1) $y_{ss} \simeq -1.4 \cos 10t$.
(2) 42, 85.9 radians/second.
(3) Yes, steady-state current has amplitude $\simeq 6.25$ amps.
(4) $T \simeq 0.49$.

CHAPTER 4

Piecewise continuous functions

4.1 INTRODUCTION

In earlier chapters, input functions, or alternatively functions on the right-hand side of differential equations, have mostly been restricted to those which can be described by a simple continuous function for $t \geqslant 0$, for example, $x = t$, $x = \sin \omega t$, etc. In this chapter we consider inputs which require a multiple definition and may have discontinuities. For example the *unit pulse function* $p_\alpha(t)$ is defined as

$$p_\alpha(t) = \begin{cases} 1 \text{ when } 0 \leqslant t < \alpha \\ 0 \text{ when } t \geqslant \alpha \end{cases},$$

and the function defined by

$$x(t) = \begin{cases} t \text{ when } 0 \leqslant t < 1 \\ 1 \text{ when } t \geqslant 1 \end{cases}$$

is a *limited* form of the ramp function.

Laplace transform techniques will be applied to this type of input, and considerable use will be made of the *shift theorem*. The concepts of pulse functions, delayed functions, and impulse functions mentioned briefly in Chapter 1 will be studied here in more detail. One example deals with a sequence of narrow width pulses and serves as an introduction to sampling devices which are considered in Chapter 5.

4.2 THE DELAY OPERATION

We define the *delayed* unit step function, $u(t-T)$, as follows:

$$u(t-T) = \begin{cases} 1 \text{ when } t \geqslant T \\ 0 \text{ when } t < T \end{cases}.$$

Delayed versions of other functions defined for $t \geqslant 0$ can be expressed in terms of $u(t-T)$.

The function $f(t-T)u(t-T)$ represents $f(t)$ delayed by amount T, that is, shifted T units to the right along the horizontal axis. Fig. 4.1 shows the delayed ramp function $(t-T)u(t-T)$.

The unit impulse function $\delta(t)$ and the delayed unit impulse function $\delta(t-T)$

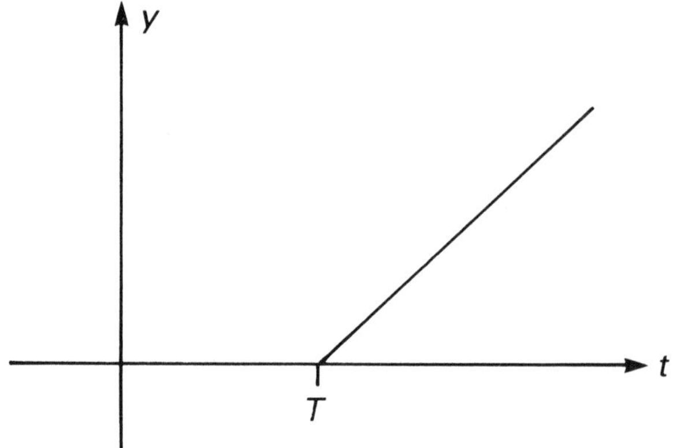

Fig. 4.1 The function $y = (t-T)\, u\, (t-T)$.

have already been defined in Chapter 1. Here it is convenient to redefine them in terms of the *unit pulse function* $p_\alpha(t)$. Thus

$$p_\alpha(t) = \begin{cases} 1 \text{ when } 0 \leqslant t < \alpha \\ 0 \text{ otherwise} \end{cases},$$

$$\delta(t) = \lim_{\alpha \to 0} \frac{p_\alpha(t)}{\alpha},$$

$$\delta(t-T) = \lim_{\alpha \to 0} \frac{p_\alpha(t-T)}{\alpha}.$$

Note that $p_\alpha(t)$ and $p_\alpha(t-T)$ can be conveniently defined in terms of step functions as follows:

$$p_\alpha(t) = u(t) - u(t-\alpha)$$

$$p_\alpha(t-T) = u(t-T) - u(t-T-\alpha).$$

From the above definitions and the definition of the Laplace transform it is easy to verify that

$$L\{u(t-T)\} = \frac{e^{-sT}}{s},$$

$$L\{p_\alpha(t)\} = \frac{1}{s}(1 - e^{-s\alpha}),$$

$$L\{p_\alpha(t-T)\} = \frac{e^{-sT}}{s}(1 - e^{-s\alpha}),$$

$$L\{\delta(t)\} = 1,$$

$$L\{\delta(t-T)\} = e^{-sT}.$$

Some of the above results are special cases of the following theorem.

Theorem 4.1 (Shift Theorem)

$$L\{f(t-T)u(t-T)\} = e^{-sT}L\{f(t)\}$$

Proof

$$L\{f(t-T)u(t-T)\} = \int_{T}^{\infty} f(t-T)e^{-st}\,dt$$

$$= \int_{0}^{\infty} f(\tau)e^{-s(\tau+T)}\,d\tau, \quad (\tau = t-T)$$

$$= e^{-sT}\int_{0}^{\infty} f(\tau)e^{-s\tau}\,d\tau$$

$$= e^{-sT}L\{f(t)\}.$$

The shift theorem is most useful in its *inverse* form,

$$L^{-1}\{e^{-sT}F(s)\} = f(t-T)u(t-T), \text{ where } F(s) = L\{f(t)\}.$$

It follows from Theorem 4.1 that the transform of a delayed function is obtained by multiplying the transform of the original function by e^{-sT}. The term e^{-sT} can be thought of as the *transfer function* of a system which has the effect of delaying its input by a time T, and is often referred to as the **delay operation**. For example, if the ramp function $f(t) = t$ is input to the system $G(s) = e^{-sT}$, then the output would be $L^{-1}\left\{\dfrac{e^{-sT}}{s^2}\right\} = (t-T)u(t-T)$, the *delayed* ramp function.

Example 4.1

The system with transfer function $G_1(s) = e^{-sT}$ has input $x(t) = a\sin\omega t$.

Write down the system output. What is the output of the system $G(s) = \dfrac{e^{-sT}}{1+s}$ when the input is $a\sin\omega t$?

The Laplace transform of the output of $G_1(s)$ is $\dfrac{a\omega e^{-sT}}{s^2 + \omega^2}$.

It follows from the inverse form of the shift theorem that the system output is $a\sin\omega(t-T)u(t-T)$.

Note that for $t < T$ the output is zero and for $t > T$ the output is $a\sin\omega(t-T)$. This 'steady-state' output can be obtained from the frequency response function $e^{-j\omega T}$ for the system $G_1(s)$. $|G_1(j\omega)| = 1$ and $\lfloor G(j\omega) = -\omega T$, and therefore the steady-state output is $a\sin(\omega t - \omega T)$.

For the system $G(s) = \dfrac{e^{-sT}}{1+s}$ the output $y(t)$ is obtained from

$$L\{y(t)\} = e^{-sT}\left(\frac{a\omega}{(s+1)(s^2+\omega^2)}\right)$$

$$= \frac{ae^{-sT}}{1+\omega^2}\left(\frac{\omega}{s+1} - \frac{\omega s - \omega}{s^2+\omega^2}\right).$$

Therefore

$$y(t) = \frac{a}{1+\omega^2}\left(\omega e^{-(t-T)} - \omega\cos\omega(t-T) + \sin\omega(t-T)\right)u(t-T).$$

That is,

$$y(t) = \begin{cases} 0 \text{ when } t < T \\ \dfrac{a}{1+\omega^2}\left(\omega e^{-(t-T)} - \omega\cos\omega(t-T) + \sin\omega(t-T)\right) \text{ when } t \geqslant T. \end{cases}$$

Example 4.2
Express the function shown in Fig. 4.2 in terms of the ramp function and a delayed function, and hence write down its Laplace transform.

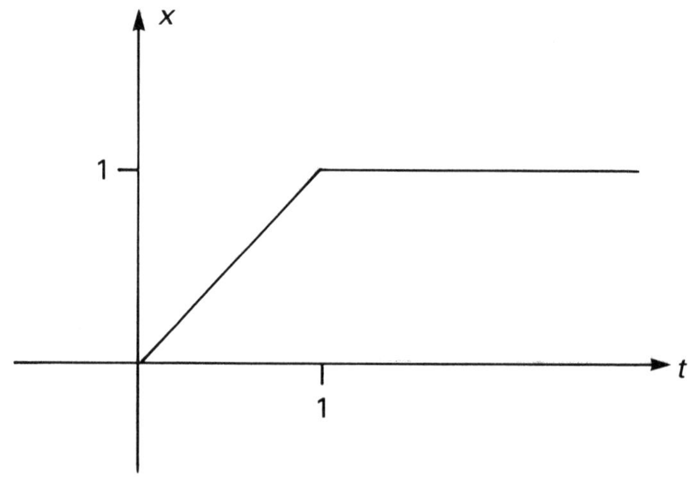

Fig. 4.2

For $0 \leqslant t < 1$, the function is $x(t) = t$. For $t \geqslant 1$, $x(t) = 1 = t - (t-1)$.

Therefore $x(t) = t - (t-1)u(t-1)$ and $L\{x(t)\} = \dfrac{1}{s^2} - \dfrac{1}{s^2} e^{-sT}$.

This result can be confirmed directly by using the definition of the Laplace transform.

Problems

(1) Find the Laplace transform of the function

(i) $\cos\left(t - \dfrac{2\pi}{3}\right) u\left(t - \dfrac{2\pi}{3}\right)$.

(ii) $f(t) = \begin{cases} 0 \text{ when } t < a \\ t - a \text{ when } a \leqslant t < b. \\ b - a \text{ when } t \geqslant b \end{cases}$

(2) For the function $f(t)$ defined in Problem 1(ii) write down the transform of $f'(t)$ and show that

$$\lim_{b \to a}\left(\frac{L\{f'(t)\}}{(b-a)}\right) = e^{-as}. \quad \text{Explain.}$$

(3) Obtain the inverse Laplace transform of

(i) $\dfrac{e^{-2s}}{s(s+1)}$.

(ii) $(1 - e^{-sT})\left(\dfrac{1}{s} - \dfrac{1}{(s+1)^2}\right)$.

(iii) $e^{-s}\left(\dfrac{e^{-2s}}{s^2} - \dfrac{e^{-4s}}{s^2}\right)$.

(4) The function $x(t)$ is defined as follows:

$$x(t) = \begin{cases} t \text{ for } 0 \leqslant t < 1 \\ 0 \text{ for } t \geqslant 1. \end{cases}$$

Using delayed step functions to express $x(t)$, write down its Laplace transform.

Find the output of the system $G(s) = \dfrac{1}{1+0.1s}$ when the input is $x(t)$.

(5) Find the Laplace transform of the staircase function $f(t) = n+1$ when $nT \leqslant t < (n+1)T$, $n = 0, 1, 2, \ldots$. If $f(t)$ is input to the system $G(s) = \dfrac{1}{1+s}$ obtain the output in the interval $nT \leqslant t < (n+1)T$.

Answers to problems

(1) (i) $\dfrac{e^{-2\pi/3\,s}}{s^2+1}$.

   (ii) $\dfrac{1}{s^2}\,(e^{-as}-e^{-bs})$.

(2) $\dfrac{1}{s}(e^{-as}-e^{-bs})$. $\dfrac{f'(t)}{b-a}$ is a unit area pulse function which tends to $\delta(t-a)$

   as $b\to a$.

(3) (i) $(1-e^{-(t-2)})u(t-2)$.

   (ii) $f(t)=\begin{cases}1-te^{-t} & \text{when } 0\leqslant t<T\\(t-T)e^{-(t-T)}-te^{-t} & \text{when } t\geqslant T\end{cases}$.

   (iii) $(t-3)u(t-3)-(t-5)u(t-5)=\begin{cases}0 & \text{when } 0\leqslant t<3\\t-3 & \text{when } 3\leqslant t<5.\\2 & \text{when } t\geqslant 5\end{cases}$

(4) $x(t)=t-(t-1)u(t-1)-u(t-1),\ L\{x(t)\}=\dfrac{1}{s^2}(1-e^{-s})-\dfrac{e^{-s}}{s},$

$t-\dfrac{1}{10}+\dfrac{1}{10}e^{-10t}-\left(t-\dfrac{1}{10}-\dfrac{9}{10}e^{-10(t-1)}\right)u(t-1).$

(5) $\dfrac{1}{s}(1+e^{-sT}+e^{-s2T}+e^{-s3T}+\ \ldots),\ \displaystyle\sum_{i=0}^{n}(1-e^{-(t-iT)})$.

4.3 PERIODIC FUNCTIONS

The periodic function $f_p(t)$ defined for $t\geqslant 0$ by

$$f_p(t)=f(t)\text{ for }0\leqslant t<L,\,f_p(t+L)=f_p(t),$$

can easily be expressed in terms of delayed step functions as follows:

$$f_p(t)=\sum_{n=0}^{\infty}f(t-nL)(u(t-nL)-u(t-[n+1]L)).$$

Hence the Laplace transform of $f_p(t)$ is

$$L\{f_p(t)\}=\sum_{n=0}^{\infty}\int_{nL}^{(n+1)L}f(t-nL)e^{-st}\,dt$$

$$=\sum_{n=0}^{\infty}\int_{0}^{L}f(\tau)e^{-s(\tau+nL)}\,d\tau$$

$$=\left(\int_{0}^{L}f(\tau)e^{-s\tau}\,d\tau\right)\sum_{n=0}^{\infty}e^{-snL}.\qquad(4.1)$$

Theorem 4.2
The Laplace transform of the periodic function $f_p(t)$ of period L is given by

$$L\{f_p(t)\} = \frac{1}{1-e^{-sL}} \int_0^L f_p(t)e^{-st}\,dt.$$

Proof
The proof follows from equation (4.1) by summing the geometric series (assuming that the real part of s is positive) and noting that $f_p(t) = f(t)$ for $0 \leqslant t < L$.

Example 4.3
Find the Laplace transform of the half-wave rectified signal

$$f(t) = \begin{cases} \sin t & \text{when } 0 \leqslant t < \pi \\ 0 & \text{when } \pi \leqslant t < 2\pi \end{cases}, \quad f_p(t+2\pi) = f_p(t).$$

From Theorem 4.2

$$L\{f_p(t)\} = \frac{1}{1-e^{-s2\pi}} \int_0^{2\pi} f_p(t)e^{-st}\,dt$$

$$= \frac{1}{1-e^{-s2\pi}} \int_0^{\pi} \sin t\, e^{-st}\,dt$$

$$= \frac{1}{1-e^{-s2\pi}} \left(\frac{1+e^{-s\pi}}{s^2+1}\right)$$

$$= \frac{1}{(1-e^{-s\pi})(s^2+1)}.$$

Note that this Laplace transform can be expressed as an infinite series

$$L\{f_p(t)\} = (1+e^{-s\pi}+e^{-s2\pi}+e^{-s3\pi}+ \ldots)\frac{1}{1+s^2},$$

and using Theorem 4.1 we have

$$f_p(t) = \sin t + \sin(t-\pi)u(t-\pi) + \sin(t-2\pi)u(t-2\pi) + \ldots$$

$$= \sin t(1 - u(t-\pi) + u(t-2\pi) - u(t-3\pi) + \ldots)$$

as an alternative representation of the given function.

Example 4.4
Find the Laplace transform of the square wave function

$$x_p(t) = \begin{cases} 1 & \text{when } 0 \leqslant t < L \\ 0 & \text{when } L \leqslant t < 2L \end{cases}, \quad x_p(t+2L) = x_p(t).$$

If $x_p(t)$ is the input to a system with transfer function $G(s) = \dfrac{1}{1+sT}$, obtain the system output $y(t)$ and show that when $L \gg T$ the *steady-state response* for the interval $2nL \leqslant t < (2n+1)L$ is

$$y_{ss}(t) = 1 - e^{-(t-2nL)/T}.$$

$$\begin{aligned}
L\{x_p(t)\} &= \frac{1}{1-e^{-s2L}} \int_0^L e^{-st}\,dt \\[2mm]
&= \frac{1}{1-e^{-s2L}} \left(\frac{1-e^{-sL}}{s} \right) \\[2mm]
&= \frac{1}{(1+e^{-sL})s} \\[2mm]
&= (1 - e^{-sL} + e^{-s2L} - e^{-s3L} + \ldots)\frac{1}{s}.
\end{aligned}$$

(This result can be obtained directly by simply expressing $x_p(t)$ in terms of delayed step functions).

The output from the system $G(s)$ is given by

$$\begin{aligned}
L\{y(t)\} &= (1 - e^{-sL} + e^{-s2L} - e^{-s3L} + \ldots)\frac{1}{s(1+sT)} \\[2mm]
&= (1 - e^{-sL} + e^{-s2L} - e^{-s3L} + \ldots)\left(\frac{1}{s} - \frac{T}{1+sT} \right).
\end{aligned}$$

Therefore $y(t) = 1 - e^{-t/T} + \displaystyle\sum_{r=1}^{\infty} (-1)^r (1 - e^{-(t-rL)/T})u(t-rL).$

In the interval $2nL \leqslant t < (2n+1)L$ we have

$$\begin{aligned}
y(t) &= 1 - \sum_{r=0}^{2n} (-1)^r e^{-(t-rL)/T} \\[2mm]
&= 1 - e^{-(t-2nL)/T} \sum_{r=0}^{2n} (-1)^r e^{(r-2n)L/T} \\[2mm]
&= 1 - e^{-(t-2nL)/T} \left(1 - \frac{e^{-L/T}(1 - e^{-2nL/T})}{1 + e^{-L/T}} \right).
\end{aligned}$$

The steady-state solution is obtained by allowing $n \to \infty$, that is,

$$y_{ss}(t) = 1 - \frac{e^{-(t-2nL)/T}}{1+e^{-L/T}}, \qquad \text{for } 2nL \leqslant t < (2n+1)L.$$

For the case $L \gg T$ we have

$$y_{ss}(t) \simeq 1 - e^{-(t-2nL)/T}, \quad 2nL \leqslant t < (2n+1)L.$$

In particular, for $n = 0$, $y_{ss} = 1 - e^{-t/T}, 0 \leqslant t < L$, which is the response of $G(s)$ to a *unit step input*. Thus this example provides an alternative proof for the technique of section 3.5.

Problems
(1) Find the Laplace transform of the sawtooth waveform defined by the function $f(t)$,

$$f(t) = \frac{kt}{T}, 0 \leqslant t < T, f(t+T) = f(t).$$

(2) Show that the Laplace transform of the full rectified wave, $f(t) = |\cos t|$, is given by

$$F(s) = \frac{s + \left(\sinh \dfrac{\pi s}{2}\right)^{-1}}{s^2 + 1}.$$

(3) Obtain the Laplace transform of the function

$$f(t) = \begin{cases} 1 \text{ when } 0 \leqslant t < L/2 \\ -1 \text{ when } L/2 \leqslant t < L \end{cases}, \quad f(t+L) = f(t).$$

Hence write down the Laplace transform of the triangular waveform

$$v(t) = \begin{cases} t \text{ when } 0 \leqslant t < L/2 \\ L-t \text{ when } L/2 \leqslant t < L \end{cases}, \quad f(t+L) = f(t).$$

Answers to problems

(1) $\dfrac{k}{Ts^2} - \dfrac{ke^{-sT}}{s(1 - e^{-sT})}.$

(3) $F(s) = \dfrac{1}{s} \tanh \dfrac{Ls}{4}, \quad V(s) = \dfrac{1}{s^2} \tanh \dfrac{Ls}{4}.$

4.4 SOLUTION OF DIFFERENTIAL EQUATIONS WITH PIECEWISE CONTINUOUS INPUTS

The output of a system for a given piecewise continuous input has already been considered briefly in section 4.2 (Problems 4 and 5) and section 4.3. In this section a number of examples are presented where the system (or differential equation) input is a piecewise continuous function. The input is expressed in terms of delayed step functions or impulses so that its Laplace transform is readily obtained.

Example 4.5

Solve the initial value problem

$$\ddot{y} + 4y = 4u(t-2),\ y(0) = 10,\ \dot{y}(0) = 0.$$

Taking Laplace transforms gives

$$s^2 L\{y\} - s10 + 4L\{y\} = \frac{4e^{-2s}}{s}.$$

Therefore $L\{y\} = \dfrac{10s}{s^2+4} + \dfrac{4e^{-2s}}{s(s^2+4)}$

$$= \frac{10s}{s^2+4} + \left(\frac{1}{s} - \frac{s}{s^2+4}\right) e^{-2s}.$$

Therefore, using Theorem 4.1,

$$y = 10\cos 2t + (1 - \cos 2(t-2))\,u(t-2),$$

or

$$y = \begin{cases} 10\cos 2t & \text{when } t < 2 \\ 10\cos 2t + 1 - \cos 2(t-2) & \text{when } t \geqslant 2. \end{cases}$$

Example 4.6

The motion of a simple mechanical system is modelled by the differential equation

$$\ddot{y} + 2\dot{y} + y = x(t),\ \ y(0) = \dot{y}(0) = 0$$

where the input function $x(t)$ is the limited ramp function defined by

$$x(t) = \begin{cases} t & \text{when } 0 \leqslant t < 1 \\ 1 & \text{when } t \geqslant 1 \end{cases}.$$

Obtain the function $y(t)$ for $t \geqslant 1$.

$$x(t) = t - (t-1)\,u(t-1),\ \text{and}$$

$$L\{x(t)\} = \frac{1}{s^2}(1 - e^{-s}).$$

Transforming the differential equation gives

$$L\{y\} = \frac{1}{s^2(s+1)^2}(1-e^{-s})$$

$$= \left(\frac{1}{s^2} - \frac{2}{s} + \frac{1}{(s+1)^2} + \frac{2}{(s+1)}\right)(1-e^{-s}).$$

Therefore

$$y(t) = t - 2 + te^{-t} + 2e^{-t} - (t - 3 + (t-1)e^{-(t-1)} + 2e^{-(t-1)})u(t-1),$$

and for $t \geqslant 1$

$$y(t) = 1 + (t+2)e^{-t} - (t+1)e^{-(t-1)}.$$

Example 4.7

Solve $\dot{y} + y = Kp_\alpha(t)$, $y(0) = 0$, where $p_\alpha(t)$ is the unit pulse function and K is a constant. Discuss the form of the solution when α is small, and the result of approximating $Kp_\alpha(t)$ by a suitable impulse function.

Transforming $\dot{y} + y = Kp_\alpha(t)$, $y(0) = 0$, we obtain

$$s L\{y\} + L\{y\} = K\left(\frac{1-e^{-s\alpha}}{s}\right), \quad \text{(see section 4.2)}.$$

Therefore $L\{y\} = \dfrac{K}{s(s+1)} - \dfrac{Ke^{-s\alpha}}{s(s+1)}$

$$= K\left(\frac{1}{s} - \frac{1}{(s+1)}\right) - Ke^{-s\alpha}\left(\frac{1}{s} - \frac{1}{(s+1)}\right).$$

Therefore $\quad y = K(1 - e^{-t}) - K(1 - e^{-(t-\alpha)})u(t-\alpha),$

that is, $\quad y = \begin{cases} K(1 - e^{-t}) \text{ for } 0 \leqslant t < \alpha \\ K(e^{-(t-\alpha)} - e^{-t}) \text{ for } t \geqslant \alpha \end{cases}.$

Fig. 4.3 shows the graph of the solution with $K(1-e^{-\alpha})$ as the peak value of y.

If α is small, then the peak value is approximately $K\alpha$ $\Bigg($ since

$1-e^{-\alpha} = 1 - 1 + \alpha - \dfrac{\alpha^2}{2} + \dfrac{\alpha^3}{6} - \ldots \Bigg)$, and $K\alpha$ is the 'area' of the pulse function $Kp_\alpha(t)$. If the pulse function had been $2Kp_{0.5\alpha}(t)$ the area is still $K\alpha$, and the peak value of y for small α is still approximately $K\alpha$, see Figs 4.4(a) and 4.4(b).

When α is small it is clear that the important parameter is the *area* of the pulse and not its height. Thus for small α we could approximate to $Kp_\alpha(t)$ using $K\alpha\delta(t)$, since the area of the impulse $\delta(t)$ is 1 by definition.

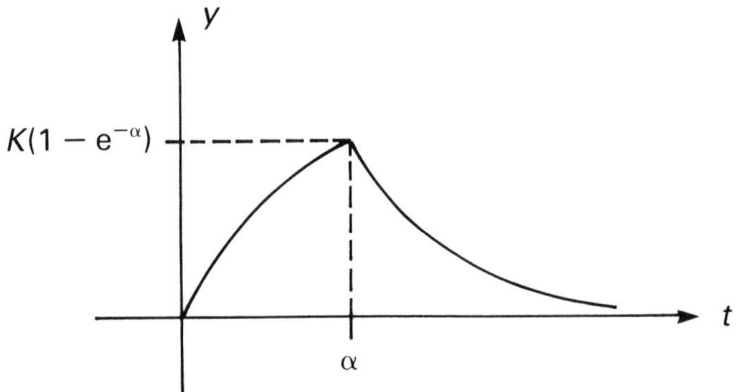

Fig. 4.3

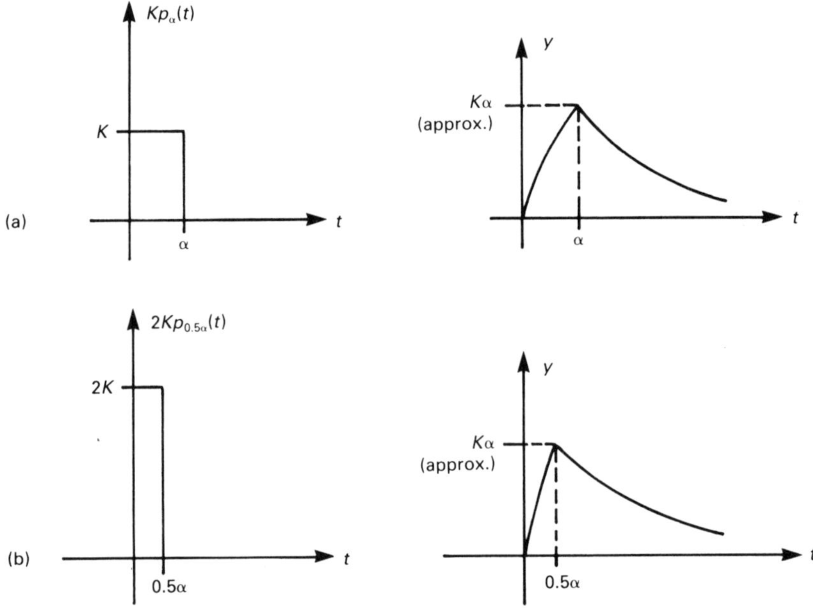

Fig. 4.4

Solving $\dot{y} + y = K\alpha\delta(t)$, $y(0) = 0$, we obtain after taking Laplace transforms

$$L\{y\} = \frac{K\alpha}{s+1}.$$

Therefore $y = K\alpha e^{-t}$ is the approximate solution (see Fig. 4.5).

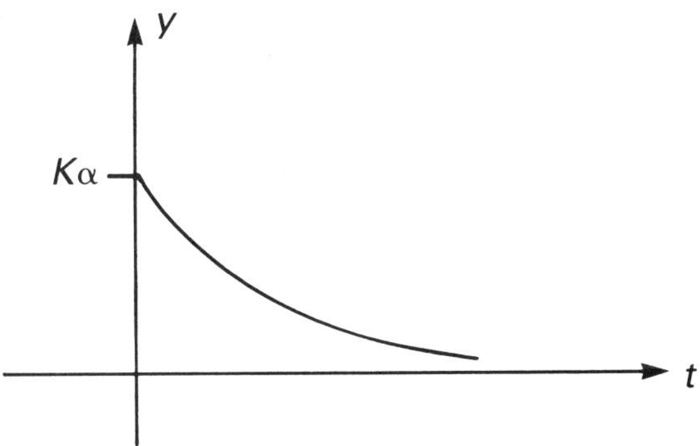

Fig. 4.5

Comparison of Figs 4.4(a) and 4.5 shows that the $\alpha\delta(t)$ approximation for $p_\alpha(t)$ is good for small α. The advantage of using $\delta(t)$ is that its Laplace transform is 1, which reduces the amount of algebraic manipulation required to complete the solution.

Example 4.8
Solve $\dot{y} + y = f(t), \quad y(0) = 0$

where $f(t) = \begin{cases} 10^4 \text{ when } 0 \leqslant t < 0.001 \\ 0 \text{ when } t \geqslant 0.001 \end{cases}$

Since the duration of the pulse is small we can use the $\delta(t)$ approximation. The 'area' of the pulse is $10^4 \times 0.001 = 10$, and we solve

$$\dot{y} + y = 10\delta(t), \ y(0) = 0.$$

Therefore $L\{y\} = \dfrac{10}{s+1}$ and $y = 10e^{-t}$, approximately.

(The reader is invited to solve the given equation without approximation and to carefully compare the two solutions by plotting accurate graphs).

Example 4.9
Consider the electronic system shown in Fig. 4.6. The differential equation modelling this system is

$$\frac{dq}{dt} + \frac{q}{RC} = \begin{cases} 0 \text{ when switch at U} \\ \dfrac{E}{R} \text{ when switch at D} \end{cases}$$

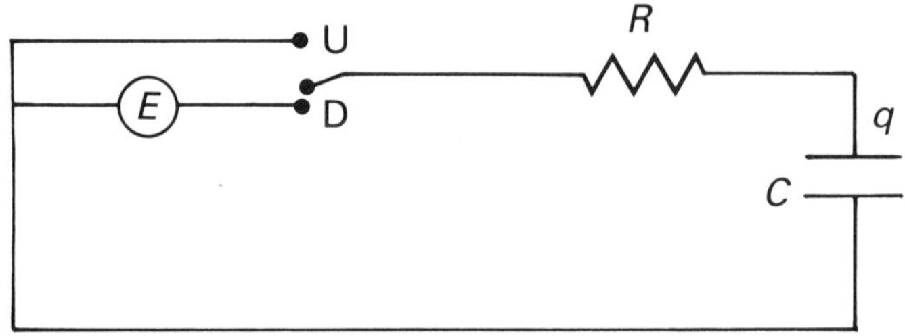

Fig. 4.6

where q is the charge on the capacitor, and E is a constant voltage. Assuming that $q(0) = 0$, solve the differential equation for each of the following situations.

(a) switch set to D at $t = 0$,
(b) switch set to D at $t = 0$ and returned to U at $t = 2$,
(c) switch set to D at $t = 0$ and returned to U at $t = \alpha$ where α is small,
(d) switch set to D for short duration α at the times $t = 0, T, 2T, 3T, \ldots$ ..

We write the differential equations as

$$\dot{q} + \frac{q}{RC} = f(t) \quad \text{and} \quad \text{interpret } f(t)$$

for the four cases.

(a) $f(t) = \frac{E}{R} u(t)$ and $L\{f(t)\} = \frac{E}{Rs}$.

Therefore $\left(s + \frac{1}{RC} \right) L\{q\} = \frac{E}{Rs}$

that is, $L\{q\} = \frac{E}{R} \left(\dfrac{1}{s\left(s + \dfrac{1}{RC} \right)} \right) = \frac{E}{R} \left(\dfrac{RC}{s} - \dfrac{RC}{\left(s + \dfrac{1}{RC} \right)} \right)$.

Therefore $q = EC (1 - e^{-t/RC})$.

(b) $f(t) = \frac{E}{R} (u(t) - u(t - 2))$ and $L\{f(t)\} = \frac{E}{Rs} (1 - e^{-2s})$.

Therefore $L\{q\} = \frac{E}{R} \left(\dfrac{1}{s\left(s + \dfrac{1}{RC} \right)} - \dfrac{e^{-2s}}{s\left(s + \dfrac{1}{RC} \right)} \right)$

and $q = EC(1 - e^{-t/RC}) - EC(1 - e^{-(t-2)/RC})u(t-2)$,

that is, $q = \begin{cases} EC(1 - e^{-t/RC}) & \text{when } 0 \leqslant t < 2 \\ EC(e^{-(t-2)/RC} - e^{-t/RC}) & \text{when } t \geqslant 2. \end{cases}$

(c) $f(t) = \dfrac{E\alpha}{R}\delta(t)$ approximately when α is small

and $L\{f(t)\} = \dfrac{E\alpha}{R}$.

Therefore $L\{q\} = \dfrac{E\alpha}{R}\left(\dfrac{1}{s + \dfrac{1}{RC}}\right)$

and $q = \dfrac{E\alpha}{R}e^{-t/RC}$.

(d) $f(t) = \dfrac{E\alpha}{R}(\delta(t) + \delta(t-T) + \delta(t-2T) + \delta(t-3T) + \ldots)$

$$\text{approximately for small } \alpha,$$

and $L\{f(t)\} = \dfrac{E\alpha}{R}(1 + e^{-sT} + e^{-s2T} + e^{-s3T} + \ldots)$.

Alternatively, $L\{f(t)\} = \dfrac{1}{1 - e^{-sT}}\displaystyle\int_0^T \dfrac{E\alpha}{R}\delta(t)\,dt$ by Theorem 4.2,

since $f(t)$ is periodic. That is,

$$L\{f(t)\} = \frac{E\alpha/R}{1 - e^{-sT}} = \frac{E\alpha}{R}(1 + e^{-sT} + e^{-s2T} + \ldots)$$

Therefore $L\{q\} = \dfrac{E\alpha}{R\left(s + \dfrac{1}{RC}\right)}(1 + e^{-sT} + e^{-s2T} + e^{-s3T} + \ldots)$

and $q = \dfrac{E\alpha}{R}(e^{-t/RC} + e^{-(t-T)/RC}u(t-T) + e^{-(t-2T)/RC}u(t-2T) + \ldots)$,

that is, $q = \begin{cases} \dfrac{E\alpha}{R}e^{-t/RC} & \text{for } 0 \leqslant t < T \\[2mm] \dfrac{E\alpha}{R}(e^{-t/RC} + e^{-(t-T)/RC}) & \text{for } T \leqslant t < 2T \\[2mm] \dfrac{E\alpha}{R}(e^{-t/RC} + e^{-(t-T)/RC} + e^{-(t-2T)/RC}) & \text{for } 2T \leqslant t < 3T, \end{cases}$

etc.

The graphs of $q \sim t$ for the four cases are shown in Fig. 4.7.

Piecewise continuous functions

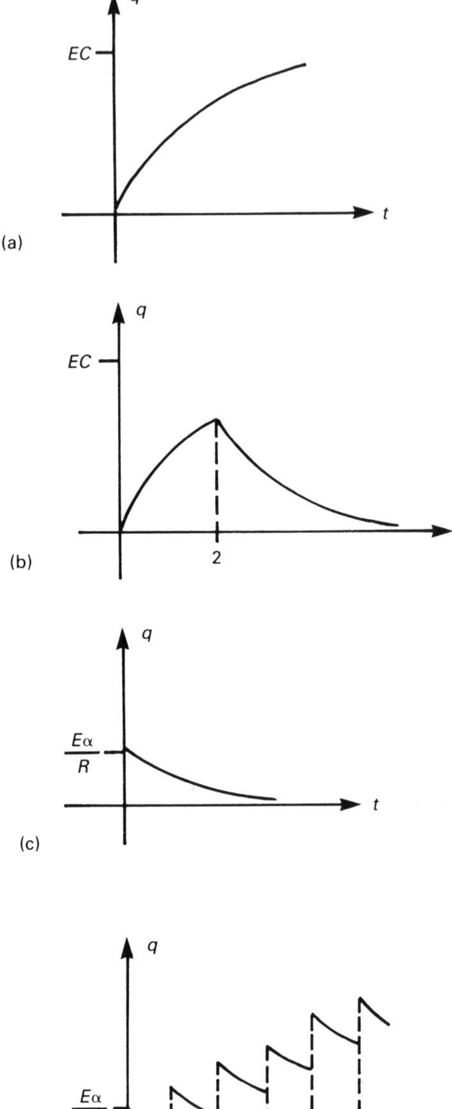

Fig. 4.7

Example 4.10
Repeat part (d) of Example 4.9 when E is replaced by

$$E \cos \frac{2\pi}{T} t.$$

Now $\dot{q} + \dfrac{q}{RC} = f(t)$ where

$$f(t) = \frac{E\alpha}{R} \left(\cos 0 \cdot \delta(t) + \cos 2\pi \cdot \delta(t-T) + \cos 4\pi \cdot \delta(t-2T) + \ldots \right)$$

approximately. Therefore $f(t) = \dfrac{E\alpha}{R}(\delta(t) + \delta(t-T) + \delta(t-2T) + \ldots)$, which is the *same function* as in Example 4.9(d). Thus the capacitor charge q is identical to that given in Example 4.9(d). The reason for this is that the switch is operating as a *sampling device* on the function $E \cos \dfrac{2\pi}{T} t$. The only significant values of this function are those at $t = 0, T, 2T, 3T, \ldots$ and all of these are equal to E. Thus so far as the sampling device is concerned the function being sampled is effectively a constant.

Example 4.11
Find the output of the system $G(s) = \dfrac{1 - e^{-s}}{s}$ when the input is $f(t)$ defined by

$$f(t) = \begin{cases} 2t \text{ when } 0 \leqslant t < 3 \\ 2 \text{ when } t \geqslant 3 \end{cases}.$$

Sketch the input and output functions for $0 \leqslant t < 5$.

$$f(t) = 2t - 2(t-3)\, u(t-3) - 4\, u(t-3)$$

and $L\{f(t)\} = \dfrac{2}{s^2} - \dfrac{2e^{-3s}}{s^2} - \dfrac{4e^{-3s}}{s}.$

Therefore $L\{y(t)\} = \dfrac{2}{s^3}(1 - e^{-3s})(1 - e^{-s}) - \dfrac{4e^{-3s}}{s^2}(1 - e^{-s})$

$$= \frac{2}{s^3}(1 - e^{-s} - e^{-3s} + e^{-4s}) - \frac{4}{s^2}(e^{-3s} - e^{-4s}).$$

Therefore $y(t) = t^2 - (t-1)^2 u(t-1) - (t-3)^2 u(t-3) + (t-4)^2 u(t-4)$

$$- 4(t-3) u(t-3) + 4(t-4) u(t-4).$$

that is, $y(t) = \begin{cases} t^2 & \text{when } 0 \leqslant t < 1 \\ 2t - 1 & \text{when } 1 \leqslant t < 3 \\ -t^2 + 4t + 2 & \text{when } 3 \leqslant t < 4 \\ 2 & \text{when } t \geqslant 4. \end{cases}$

The graphs of $f(t)$ and $y(t)$ are shown in Fig. 4.8. The transfer function

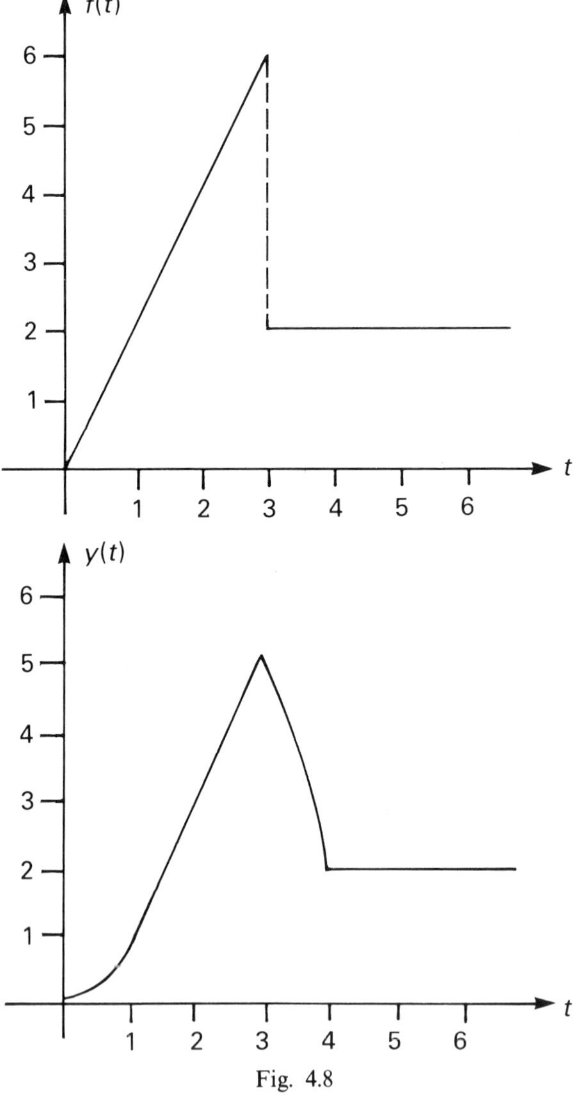

Fig. 4.8

considered in this example is a special case of $\dfrac{1-e^{-sT}}{s}$ which is important in the study of sampling techniques and reconstruction of sampled functions (see Chapter 7).

Problems

(1) Solve the differential equation

$$\ddot{y} + 2\dot{y} + y = x(t),\ y(0) = \dot{y}(0) = 0,$$

$$\text{where } x(t) = \begin{cases} t & \text{when } 0 \leqslant t < 1 \\ 0 & \text{when } t \geqslant 1 \end{cases}.$$

(2) Repeat Problem 1 when

$$\text{(i)}\quad x(t) = \begin{cases} 1 & \text{when } 0 \leqslant t < 2 \\ 0 & \text{when } t \geqslant 2 \end{cases}.$$

$$\text{(ii)}\quad x(t) = \begin{cases} 2 \times 10^5 & \text{when } 3 \leqslant t < 3.001 \\ 0 & \text{otherwise} \end{cases}.$$

(3) Solve the differential equation

$$\ddot{x} + 2\dot{x} + 5x = 10\delta(t),\ x(0) = 0,\ \dot{x}(0) = 10.$$

(4) An undamped mechanical system with natural frequency ω_n radians per second is modelled by the transfer function $G(s) = \dfrac{\omega_n^2}{s^2 + \omega_n^2}$. Its input is the periodic square wave $x_p(t)$ defined by

$$x_p(t) = \begin{cases} 1 & \text{when } 0 \leqslant t < L \\ 0 & \text{when } L \leqslant t < 2L \end{cases},\ x_p(t + 2L) = x_p(t).$$

Consideration of the Fourier series expansion of $x_p(t)$ shows that we can expect resonance whenever L is such that

$$\omega_n = \frac{(2r - 1)\pi}{L},\ r = 1, 2, 3, \ldots.$$

Confirm this by using Laplace transform techniques.

(5) A uniform beam is simply supported at its ends $x = 0$ and $x = L$. The beam has weight W and is subjected to a concentrated load P at $x = L/2$. The deflection of the beam at position x is given by the solution of the differential equation

$$EI\frac{d^4 y}{dx^4} = \frac{W}{L} + P\delta\left(x - \frac{L}{2}\right)$$

where E and I are constants. The boundary conditions for the simply supported beam are

$$y(0) = y(L) = 0, \ y''(0) = y''(L) = 0.$$

Use Laplace transform techniques to express y as a function of x.

(6) Find the output of the system $G(s) = \dfrac{1}{1+sT}$ when the input is

$$x(t) = \begin{cases} 2 - t \text{ when } 0 \leqslant t < 1 \\ 1 \text{ when } t \geqslant 1 \end{cases}.$$

(7) Find the output of the system $G(s) = \dfrac{1 - e^{-3s}}{s(s+1)}$ when the input is a unit step function.

(8) The input to the filter shown in Fig. 4.9 is given by

$$x(t) = \begin{cases} 1 \text{ when } 0 \leqslant t < 2 \\ 0 \text{ when } t \geqslant 2 \end{cases}.$$

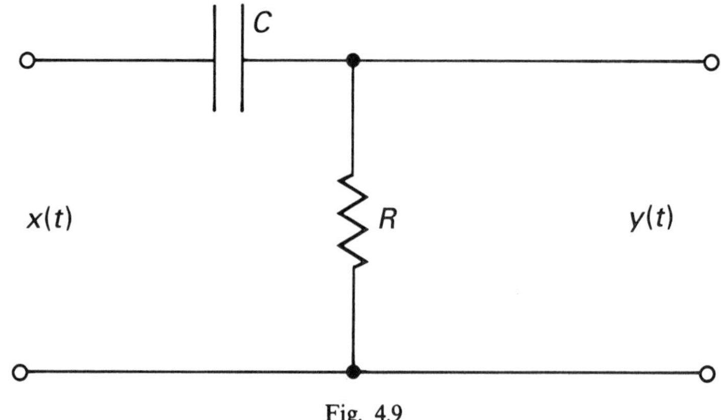

Fig. 4.9

Find the output $y(t)$ for $t \geqslant 2$.

(9) In Example 4.9(d) show that

$$\dot{q} = \frac{E\alpha}{R}\left(\frac{1 - e^{-(n+1)T/RC}}{1 - e^{-T/RC}}\right) \quad \text{when } t = nT^+.$$

What is the maximum level reached by q?

Answers to problems

(1) $y = \begin{cases} t - 2 + (t+2)e^{-t} \text{ when } 0 \leqslant t < 1 \\ (t+2)e^{-t} - e^{-(t-1)} \text{ when } t \geqslant 1 \end{cases}.$

(2) (i) $y = \begin{cases} 1 - (t+1)e^{-t} & \text{when } 0 \leqslant t < 2 \\ -(t+1)e^{-t} + (t-1)e^{-(t-2)} & \text{when } t \geqslant 2 \end{cases}$.

(ii) $y = 200(t-3)e^{-(t-3)}u(t-3)$.

(3) $x = 10e^{-t}\sin 2t$.

(5) $y = \dfrac{1}{48EI}\left(\dfrac{2W}{L}x^4 + 8P\left(x - \dfrac{L}{2}\right)^3 u\left(x - \dfrac{L}{2}\right)\right.$

$$\left. - 4(W+P)x^3 + (2W+3P)L^2 x\right).$$

(6) $y = \begin{cases} -t + (T+2)(1 - e^{-t/T}) & \text{when } 0 \leqslant t < 1 \\ 1 - (T+2)e^{-t/T} + Te^{-(t-1)/T} & \text{when } t \geqslant 1 \end{cases}$.

(7) $y = \begin{cases} t - 1 + e^{-t} & \text{when } 0 \leqslant t < 3 \\ 3 + e^{-t} - e^{-(t-3)} & \text{when } t \geqslant 3 \end{cases}$.

(8) $y(t) = e^{-t/RC} - e^{-(t-2)/RC}$ when $t \geqslant 2$.

(9) $\dfrac{E\alpha}{R}\left(\dfrac{1}{1 - e^{-T/RC}}\right)$.

Part 2
Discrete systems

CHAPTER 5

The ideal sampler and the z-transform

5.1 INTRODUCTION

In this chapter we distinguish between digital systems and sampled-data systems. A theoretical device, the ideal sampler, is defined which leads to the formulation of the z-transform. The z-transform is a useful tool in the analysis of both digital systems and sampled-data systems.

5.2 LINEAR SYSTEMS

It is necessary to distinguish the three categories of linear system shown in Fig. 5.1. In Figs 5.1(a) and (c) the system is linear when the continuous process can be described by a linear differential equation (that is, represented by a transfer function, $F(s)$).

In Fig. 5.1(b) the symbol $\{x_n\}$ represents the ordered sequence of numbers x_0, x_1, x_2, x_3, ... The system will be linear when the digital processor calculates the number y_k of the sequence $\{y_n\}$ as a linear combination of x_0, x_1, . . . x_k, y_0, y_1, . . . y_{k-1} (that is, the digital processor can be described by a linear difference equation).

The sequence $\{x_n\}$ may have arisen by taking samples (usually at equally spaced time intervals) from some continuous function $x(t)$, the sequence being called a *time series*.

The sampling device may be an ADC (Analog-to-digital-converter) which outputs the time series $\{x_n\}$. The corresponding output time series $\{y_n\}$ may or may not be reconstructed to form a continuous function, depending on the particular application.

In Fig. 5.1(c) the sampler behaves in a similar fashion to the switch in Example 4.9(d). The output of the sampled-data system is a continuous function which may require to be sampled, depending on the particular application.

Fig. 5.2 shows an example where all three types of system are present at the same time.

5.3 THE IDEAL SAMPLER

The switch sampler described in Example 4.9(d) had the effect of replacing the

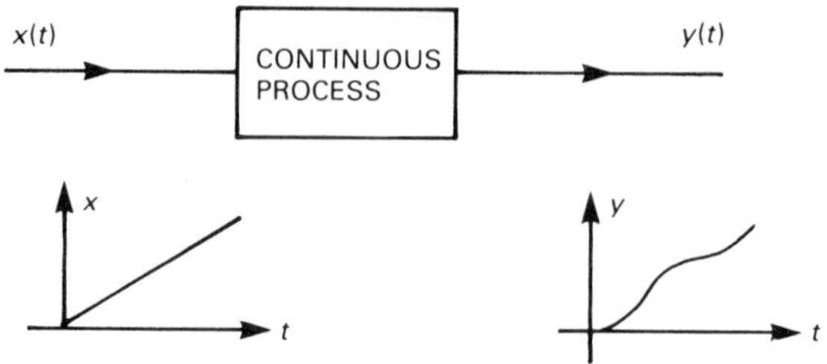

(a)　Continuous system

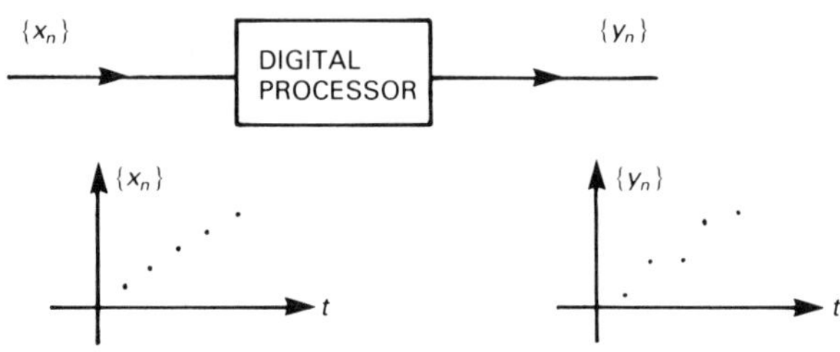

(b)　Digital system

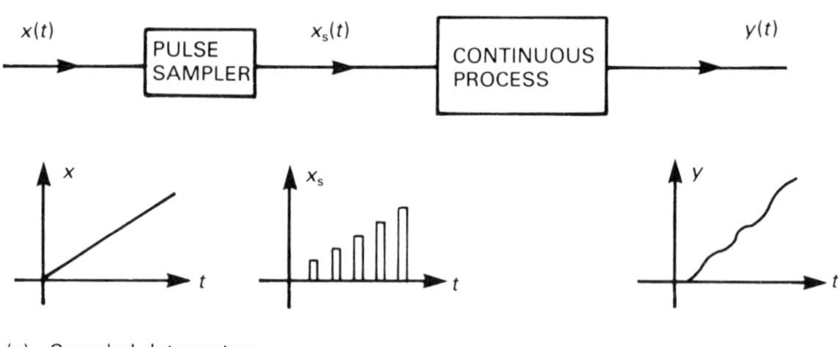

(c)　Sampled-data system

Fig. 5.1 Linear systems.

constant E/R by

$$\frac{E\alpha}{R}\left(\delta(t)+\delta(t-T)+\delta(t-2T)+\ \ldots\right),$$

where α was the closure time of the switch (assumed small). If such a sampler is applied to the function $f(t)$ the output would be represented by

$$\alpha\left(f(0)\delta(t)+f(T)\delta(t-T)+f(2T)\delta(t-2T)+\ \ldots\right).$$

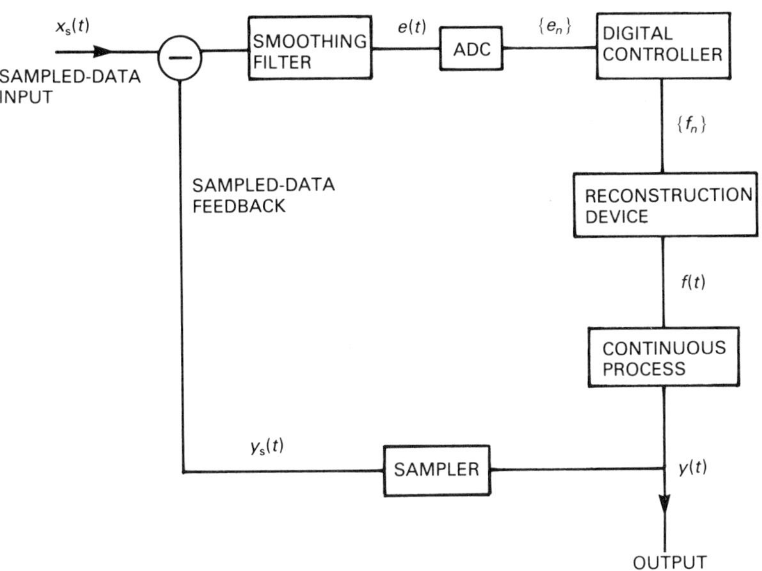

Fig. 5.2

This is not a very satisfactory device since the strength of the output signal depends on the closure time of the switch. To overcome this disadvantage it would be necessary to amplify the output signal by an amount $1/\alpha$ so that the output becomes

$$\sum_{n=0}^{\infty} f(nT)\,\delta(t-nT).$$

We define a theoretical device known as the **ideal sampler** which converts the continuous function $f(t)$ into the train of impulses

$$\sum_{n=0}^{\infty} f(nT)\,\delta(t-nT).$$

The notation $f^*(t)$ is used to denote the train of impulses, and the symbol

used for the ideal sampler is $\underset{T}{\longrightarrow}$ where T is the sampling interval.

Fig. 5.3 shows a typical situation.

The analysis of a system containing an ideal sampler may be carried out by using Laplace transforms. Example 5.1 demonstrates the technique.

A type of sampling device commonly used is the **sample/hold** device. Its output holds the value of the input until the next sampling time is reached, when the output is updated, see Fig. 5.4.

The ideal sampler and the z-transform

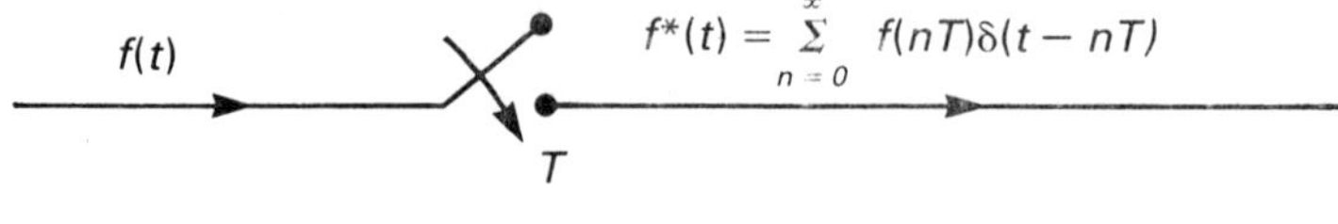

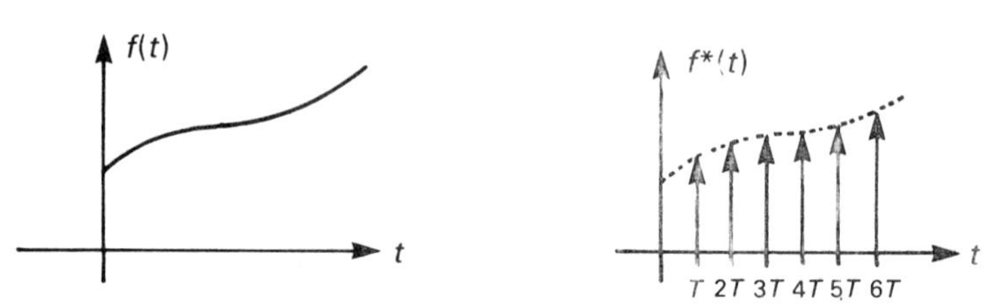

Fig. 5.3 The ideal sampler. The arrow at $t = nT$ represents an impulse of strength $f(nT)$.

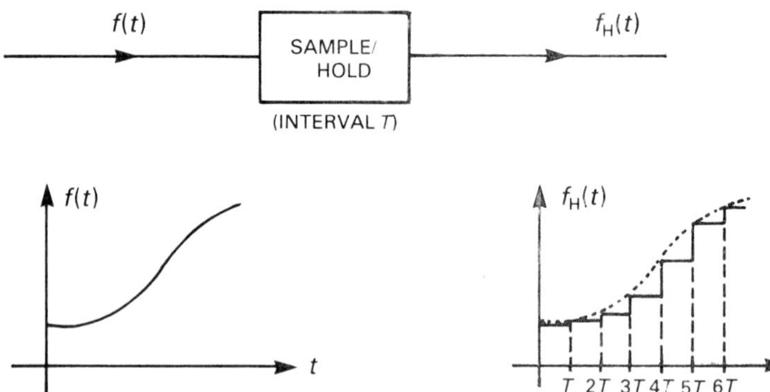

Fig. 5.4 The sample/hold device.

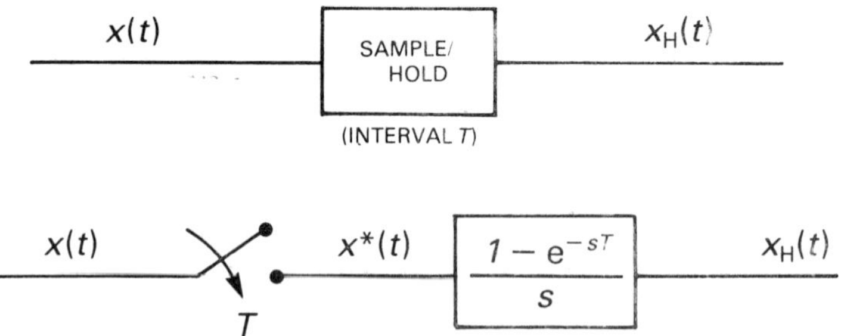

Fig. 5.5 Equivalent systems.

The analysis of a system including a sample/hold device can be carried out by consideration of an equivalent system which incorporates an ideal sampler. Fig. 5.5 shows the equivalent systems, and the proof is given in Example 5.2. Further examples are considered in Chapter 7.

Example 5.1
Find the output of the sampled-data system shown in Fig. 5.6 given that the input $x(t) = t$, $t \geq 0$.

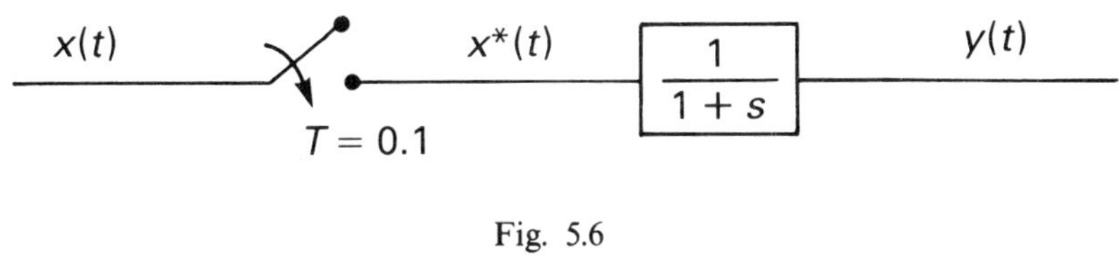

Fig. 5.6

$$x(t) = t \text{ and } x^*(t) = 0.1\delta(t - 0.1) + 0.2\delta(t - 0.2) + 0.3\delta(t - 0.3) + \ldots$$

Therefore $L\{x^*(t)\} = 0.1e^{-0.1s} + 0.2e^{-0.2s} + 0.3e^{-0.3s} + \ldots$

and $L\{y(t)\} = \dfrac{1}{1+s}(0.1e^{-0.1s} + 0.2e^{-0.2s} + 0.3e^{-0.3s} + \ldots)$.

Therefore $y(t) = 0.1e^{-(t-0.1)}u(t - 0.1) + 0.2e^{-(t-0.2)}u(t - 0.2)$
$$+ 0.3e^{-(t-0.3)}u(t - 0.3) + \ldots.$$

That is, $y(t) = 0$ for $0 \leq t < 0.1$,

and $y(t) = \displaystyle\sum_{r=1}^{n} 0.1\,re^{-(t-0.1r)}$ for $0.1n \leq t < 0.1(n+1), n = 1, 2, 3 \ldots$.

Example 5.2
Prove the equivalence of the systems shown in Fig. 5.5. Hence find the output of the system shown in Fig. 5.7 when the input is $x(t) = t$, $t \geq 0$.

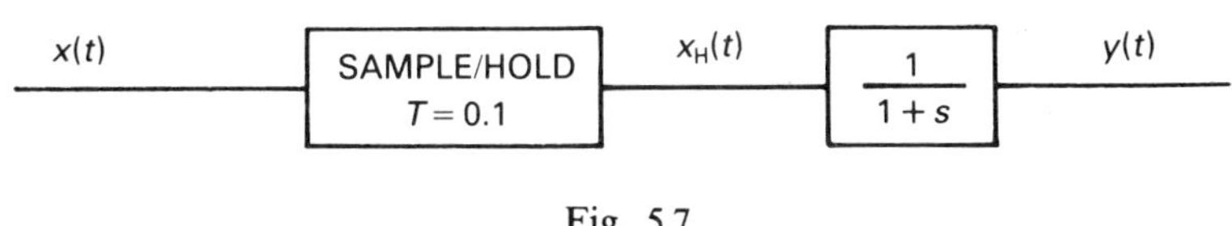

Fig. 5.7

Expressing $x_H(t)$ in terms of delayed step functions, we obtain

$$x_H(t) = x(0)u(t) + (x(T) - x(0))u(t-T) + (x(2T)) - x(T))u(t-2T) + \ldots.$$

Therefore

$$L\{x_H(t)\} = \frac{1}{s}(x(0) + (x(T) - x(0))e^{-sT} + (x(2T) - x(T))e^{-s2T} + \ldots)$$

$$= \left(\frac{1-e^{-sT}}{s}\right)(x(0) + x(T)e^{-sT} + x(2T)e^{-s2T} + \ldots)$$

$$= \left(\frac{1-e^{-sT}}{s}\right)L\{x^*(t)\},$$

which proves the equivalence.

For the system in Fig. 5.7 $x(t) = t$,

and $\qquad L\{x^*(t)\} = 0.1e^{-s0.1} + 0.2e^{-s0.2} + \ldots$ as in Example 5.1.

Therefore $\quad L\{x_H(t)\} = \left(\frac{1-e^{-s0.1}}{s}\right)(0.1e^{-s0.1} + 0.2e^{-s0.2} + \ldots)$

$$= \frac{0.1}{s}(e^{-s0.1} + e^{-s0.2} + \ldots).$$

Therefore $\quad L\{y(t)\} = \frac{0.1}{s(s+1)}(e^{-s0.1} + e^{-s0.2} + e^{-s0.3} + \ldots)$

and $y(t) = 0.1((1 - e^{-(t-0.1)})u(t-0.1) + (1 - e^{-(t-0.2)})u(t-0.2) + \ldots)$.

That is,

$$y(t) = \begin{cases} 0 \text{ when } t < 0.1 \\ 0.1 \sum_{r=1}^{n}(1 - e^{-(t-0.1r)}) \text{ when } 0.1n \leqslant t < 0.1(n+1), \ n = 1, 2, \ldots \end{cases}$$

Problems

(1) Obtain the output of the system shown in Fig. 5.8 when the input $x(t)$ is the unit step function.

(2) Find the output of the system shown in Fig. 5.9.

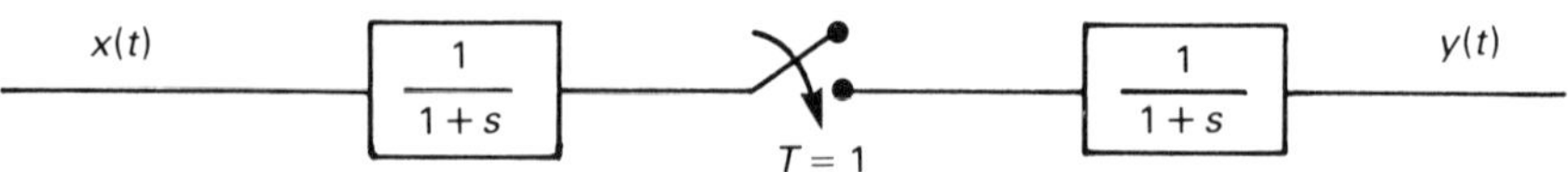

Fig. 5.8

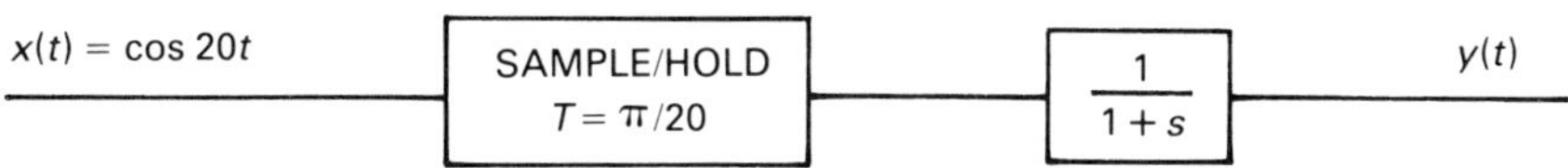

Fig. 5.9

Answers to problems

(1) $y(t) = \dfrac{e(e^n - 1)e^{-t}}{e - 1} - ne^{-t},\ n \leqslant t < n + 1,\ n = 0, 1, 2, \ldots$

(2) $y(t) = (-1)^n - e^{-t} + \dfrac{2e^T(1 - (-e^{-T})^n)e^{-t}}{1 + e^T},$

$$nT \leqslant t < (n+1)T,\ T = \frac{\pi}{20},\ n = 0, 1, 2, \ldots$$

5.4. THE z-TRANSFORM

Suppose that the sampled-data system considered in Example 5.1 has its output sampled at times 0, 0.1, 0.2, 0.3, ..., to generate the time series $\{y_n\} = \{y(0.1n)\}$. y_n can be readily obtained from the result of Example 5.1,

$$y_n = y(0.1\,n) = \sum_{r=1}^{n} 0.1\,r\,e^{-0.1(n-r)}. \tag{5.1}$$

This process can be generalized for the system shown in Fig. 5.10. Using Laplace transforms we have

$$L\{y(t)\} = F(s)L\{x^*(t)\}$$

$$= F(s) \sum_{r=0}^{\infty} x(rT)e^{-srT}.$$

Therefore

$$y(t) = L^{-1}\left\{ F(s) \sum_{r=0}^{\infty} x(rT)e^{-srT} \right\}$$

$$= \sum_{r=0}^{\infty} x(rT)f(t - rT)u(t - rT), \tag{5.2}$$

using Theorem 4.1, where $f(t) = L^{-1}\{F(s)\}$.

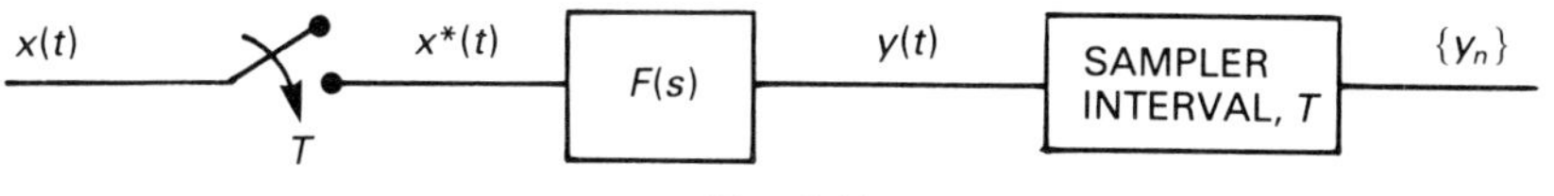

Fig. 5.10

Finally y_n is obtained from Equation (5.2) by replacing t by nT, noting that $u((n-r)T) = 0$ for $r > n$. That is,

$$y_n = y(nT)$$

$$= \sum_{r=0}^{n} x(rT) f((n-r)T) \tag{5.3}$$

(Note the *convolution* form of Equation (5.3).)

In situations where *only* the sequence $\{y_n\}$ is required and *not* the function $y(t)$, the following alternative formulation is valuable. It is based on the definition of the **z-transform** of a sequence and leads to *transfer function theory* applicable to sampled-data and digital systems.

We define the z-transform of the sequence of numbers $\{x_n\} = x_0, x_1, x_2, x_3,$ . . . as the following infinite series in *descending* powers of z. We denote the z-transform of the sequence by $Z\{x_n\}$.

$$Z\{x_n\} = \sum_{n=0}^{\infty} x_n z^{-n} = x_0 + x_1 z^{-1} + x_2 z^{-2} + \ldots \tag{5.4}$$

This series will be convergent for $|z| > R$ for some number R depending on $\{x_n\}$.

Frequently the sequence $\{x_n\}$ arises from the sampling of a function $x(t)$ at times $t = 0, T, 2T, 3T, \ldots$ That is,

$$\{x_n\} = \{x(nT)\}$$

and
$$Z\{x(nT)\} = \sum_{n=0}^{\infty} x(nT) z^{-n}.$$

This is commonly written

$$Z\{x(t)\} = \sum_{n=0}^{\infty} x(nT) z^{-n}$$

$$= x(0) + x(T) z^{-1} + x(2T) z^{-2} + \ldots \tag{5.5}$$

When writing $Z\{x(t)\}$ we must realize that the z-transformation applies *not* to $x(t)$ but to the sequence $x(0), x(T), x(2T), x(3T), \ldots$ The z-transform process has the effect of converting a *sequence* into an *infinite series*.

Note that the z-transform of the sequence $x(nT)$ is equivalent to the Laplace transform of $x^*(t)$ with e^{sT} replaced by z, since

$$L\{x^*(t)\} = \sum_{n=0}^{\infty} x(nT) e^{-snT}, \quad \text{and}$$

$$Z\{x(t)\} = \sum_{n=0}^{\infty} x(nT) z^{-n}.$$

The relationship $z = e^{sT}$ is used extensively throughout the remaining chapters.

Returning to the system of Fig. 5.10 and noting its convolution relationship (5.3), we develop the *transfer function relationship* in terms of z-transforms. (We have already seen in Chapter 1 the correspondence between the convolution integral and transfer function for a continuous system.) The result is only *demonstrated* at this stage; the proof is delayed until Chapter 7 since the main subject matter of the present chapter is to introduce the z-transform and consider some of its properties.

The transfer function form corresponding to Fig. 5.10 is

$$Z\{y(t)\} = Z\{f(t)\} \cdot Z\{x(t)\}, \tag{5.6}$$

where
$$f(t) = L^{-1}\{F(s)\}. \tag{5.7}$$

$Z\{f(t)\}$ is the transfer function of the system relating the input function $x(t)$ to the output *sequence* $\{y(nT)\}$. Denoting the transfer function by $F(z)$, the required output sequence is obtained by taking the *inverse z-transform*,

$$\{y(nT)\} = Z^{-1}\{F(z) \cdot Z\{x(t)\}\}. \tag{5.8}$$

The inverse z-transform converts an infinite series in z^{-1} to a sequence of numbers as follows

$$Z^{-1}\{y_0 + z^{-1}y_1 + z^{-2}y_2 + \dots\} = \{y_n\}. \tag{5.9}$$

It is important to understand that the *continuous function* $y(t)$ cannot be obtained by equation (5.8).

Example 5.3

Write down the transfer function corresponding to Fig. 5.6. Use z-transforms to obtain the output sequence $y(0.1n)$ for $x(t) = t$.

$$F(s) = \frac{1}{1+s}, \text{ therefore } f(t) = e^{-t}.$$

It follows that

$$F(z) = 1 + e^{-0.1}z^{-1} + e^{-0.2}z^{-2} + e^{-0.3}z^{-3} + \dots$$

is the transfer function relating $x(t)$ and $\{y(0.1n)\}$. Now $x(t) = t$, and therefore

$$Z\{x(t)\} = 0.1z^{-1} + 0.2z^{-2} + 0.3z^{-3} + \dots$$

From equation (5.8) the z-transform of $\{y(0.1n)\}$ is obtained by taking the product of these infinite series. Therefore

$$\begin{aligned}Z\{y(0.1n)\} = {}&0.1z^{-1} + (0.2 + 0.1e^{-0.1})z^{-2} \\ &+ (0.3 + 0.2e^{-0.1} + 0.1e^{-0.2})z^{-3} + \dots\end{aligned}$$

The output sequence $\{y(0.1n)\}$ is obtained by taking the inverse z-transform, that is, the set of coefficients of descending powers of z. Therefore

$$
\begin{aligned}
y(0) \ \ &= 0 \text{ (the coefficient of } z^0\text{)},\\
y(0.1) &= 0.1 \text{ (the coefficient of } z^{-1}\text{)},\\
y(0.2) &= 0.2 + 0.1e^{-0.1} \text{ (the coefficient of } z^{-2}\text{)},\\
y(0.3) &= 0.3 + 0.2e^{-0.1} + 0.1e^{-0.2} \text{ (the coefficient of } z^{-3}\text{)},
\end{aligned}
$$

etc.

It is easily seen that

$$
y(0.1n) = \sum_{r=1}^{n} 0.1r\,e^{-0.1(n-r)}
$$

as already obtained in equation (5.1).

The above example demonstrates that an inverse z-transform can be obtained simply by taking the coefficients in an infinite series. Other methods of obtaining inverse z-transforms will be considered in Chapter 6.

Problem

(1) Repeat Example 5.3 for $x(t) = 1 - e^{-t}$ and $T = 1$. Compare your answer with Problem 1 in section 5.3.

5.5 z-TRANSFORMS OF SIMPLE SEQUENCES

For many simple sequences the infinite series can be summed to obtain the z-transform in closed form. It is worth remembering the particular series

$$
\frac{1}{1-x} = 1 + x + x^2 + x^3 + \ldots, \quad |x| < 1.
$$

$$
\frac{1}{(1-x)^2} = 1 + 2x + 3x^2 + 4x^3 + \ldots, \quad |x| < 1.
$$

The following examples illustrate the technique.

(i) The sequence $x_n = 1$, $n = 0, 1, 2, 3, \ldots$.

$$
\begin{aligned}
Z\{x_n\} &= 1 + z^{-1} + z^{-2} + z^{-3} + \ldots\\
&= \frac{1}{1 - z^{-1}}\\
&= \frac{z}{z-1}.
\end{aligned}
$$

Alternatively we can write $Z\{u(t)\} = \dfrac{z}{z-1}$, where $u(t)$ is the unit step function.

(ii) The sequence $x_n = c^n$, $n = 0, 1, 2, 3, \ldots$, where c is a constant.

$$Z\{x_n\} = 1 + cz^{-1} + c^2 z^{-2} + c^3 z^{-3} + \ldots$$

$$= \frac{1}{1 - cz^{-1}}$$

$$= \frac{z}{z - c}.$$

To obtain the z-transform of the sequence
$\{x(nT)\} = \{a^{nT}\}$ we write
$a^{nT} = (a^T)^n$ and use the above result with $c = a^T$.

Therefore $Z\{a^t\} = Z\{a^{nT}\} = \dfrac{z}{z - a^T}.$

It follows that, for the sequence
$x(t) = e^{kt}$, $t = 0, T, 2T, 3T, \ldots,$

$$Z\{e^{kt}\} = \frac{z}{z - e^{kT}}.$$

(iii) The sequence $x = n$, $n = 0, 1, 2, 3, \ldots$.

$$\begin{aligned}
Z\{x_n\} &= z^{-1} + 2z^{-2} + 3z^{-3} + 4z^{-4} + \ldots \\
&= z^{-1}(1 + 2z^{-1} + 3z^{-2} + 4z^{-3} + \ldots) \\
&= z^{-1}(1 - z^{-1})^{-2}
\end{aligned}$$

$$= \frac{z}{(z - 1)^2}.$$

To obtain the z-transform of the sequence
$x(t) = t$, $t = 0, T, 2T, 3T, \ldots,$
we note that $x(nT) = nT$, $n = 0, 1, 2, 3, \ldots$

Therefore $Z\{t\} = \dfrac{Tz}{(z - 1)^2}.$

Table 2 gives the z-transforms of sequences formed by sampling certain common functions at times $t = 0, T, 2T, 3T, \ldots$. The table can be converted to one involving sequences as functions of n by the substitution $t = nT$ in the left-hand column. For example

(i) the sequence $\{nT\}$ transforms to $\dfrac{Tz}{(z - 1)^2}$

that is, for $\{x_n\} = \{n\}$ we have

$$Z\{x_n\} = \frac{z}{(z - 1)^2}.$$

The ideal sampler and the z-transform

Table 2 z-transforms of $f(t)$, $f(t) = 0$ for $t < 0$.

$$Z\{f(t)\} = \sum_{n=0}^{\infty} f(nT)z^{-n} = F(z)$$

$f(t)$, $t = 0, T, 2T, \ldots$	$F(z)$
1	$\dfrac{z}{z-1}$
t	$\dfrac{zT}{(z-1)^2}$
t^2	$\dfrac{T^2 z(z+1)}{(z-1)^3}$
$tf(t)$	$-zTF'(z)$
$u(t-T)$	$\dfrac{1}{z-1}$
$u(t) - u(t-T)$	1
$u(t-kT)$, $k = 1, 2, 3, \ldots$	$\dfrac{1}{z^{k-1}(z-1)}$
$u(t-kT) - u(t-[k+1]T)$	$\dfrac{1}{z^k}$
a^t	$\dfrac{z}{z-a^T}$
e^{ct}	$\dfrac{z}{z-\mathrm{e}^{cT}}$
$a^{t/T}$	$\dfrac{z}{z-a}$
$\dfrac{t}{T}a^{(t-T)/T}$	$\dfrac{z}{(z-a)^2}$
$a^{(t-T)/T}$	$\dfrac{1}{z-a}$
$\cos bt$	$\dfrac{z(z-\cos bT)}{z^2 - 2z\cos bT + 1}$
$\sin bt$	$\dfrac{z\sin bT}{z^2 - 2z\cos bT + 1}$
$\mathrm{e}^{ct}f(t)$	$F(\mathrm{e}^{-cT}z)$
$f(t+T)$	$zF(z) - zf(0)$

Table 2   Continued.

$f(t), t = 0, T, 2T, \ldots$	$F(z)$
$f(t + 2T)$	$z^2 F(z) - z^2 f(0) - z f(T)$
$f(t + kT)$	$z^k F(z) - z^k \sum_{n=0}^{k-1} f(nT) z^{-n}$
$f(t - kT) u(t - kT)$	$z^{-k} F(z)$

Convention for jump discontinuities
If $f(t)$ has a jump discontinuity at $t = nT$, we interpret $f(nT)$ as follows:

$$f(nT) = \lim_{t \to nT+} f(t).$$

(ii) $Z^{-1}\left\{\dfrac{z}{z - a^T}\right\} = a^{nT}, n = 0, 1, 2, 3, \ldots$

that is, $Z^{-1}\left\{\dfrac{z}{z - c}\right\} = c^n, n = 0, 1, 2, 3, \ldots$

The following properties of the z-transform follow directly from the definition

$$Z\{f(t)\} = \sum_{n=0}^{\infty} f(nT) z^{-n}.$$

The properties are useful for obtaining the z-transforms of non-trivial functions, and for dealing with difference equations (see Chapter 6).

Many of the properties correspond to results already obtained for the Laplace transform. In particular, property (ii) demonstrates that multiplication by z^{-1} has the effect of *delaying* the sequence by one sampling interval. This should not be surprising since $z^{-1} = (e^{sT})^{-1} = e^{-sT}$, and we have already seen that, for continuous functions, multiplication by e^{-sT} is equivalent to a delay of amount T, (Theorem 4.1).

Property (iii) is proved in Example 5.5. Proof of the other properties is left for the reader.

Further properties of z-transforms of the sequence $f(nT), n = 0, 1, 2, 3 \ldots$ are

(i)   (Linearity property)
$$Z\{c_1 f_1(t) + c_2 f_2(t)\} = c_1 Z\{f_1(t)\} + c_2 Z\{f_2(t)\}.$$

(ii)   (Delay operation)
$$Z\{f(t - T) u(t - T)\} = z^{-1} Z\{f(t)\}.$$

(iii) $Z\{f(t+T)\} = zZ\{f(t)\} - zf(0).$

(iv) $Z\{f(t-kT)u(t-kT)\} = z^{-k}Z\{f(t)\}, \ k = 1, 2, 3 \ldots..$

(v) $Z\{f(t+kT)\} = z^k Z\{f(t)\} - \sum_{i=0}^{k-1} z^{k-i} f(iT), \ k = 1, 2, 3, \ldots..$

(vi) $Z\{e^{ct}f(t)\} = F(e^{-cT}z)$ where $F(z) = Z\{f(t)\}.$

Example

$$Z\{u(t)\} = \frac{z}{z-1}.$$

Therefore $Z\{e^{ct}u(t)\} = \dfrac{e^{-cT}z}{e^{-cT}z-1},$

that is, $\qquad Z\{e^{ct}\} = \dfrac{z}{z-e^{cT}}.$

(vii) $Z\{tf(t)\} = -zTF'(z)$ where $F(z) = Z\{f(t)\}.$

Example

$$Z\{t\} = \frac{Tz}{(z-1)^2}.$$

Therefore $Z\{t^2\} = -zT\dfrac{d}{dz}\left(\dfrac{Tz}{(z-1)^2}\right)$

$$= -zT^2\left(\frac{(z-1)^2 - 2z(z-1)}{(z-1)^4}\right)$$

$$= \frac{T^2 z(z+1)}{(z-1)^3}.$$

(viii) (Initial value theorem)

$$f(0) = \lim_{z \to \infty} (F(z))$$

(ix) (Final value theorem)

$$\lim_{t \to \infty} (f(t)) = \lim_{z \to 1^+} ((z-1)F(z)),$$

provided the series $F(z)$ is convergent for $|z| > 1.$

Example 5.4

(a) Use the definition of the z-transform to obtain

$$Z\{te^{3t}\}, \ t = 0, T, 2T, 3T, \ldots..$$

(b) Find the sequence $\{x_n\}$ such that

$$Z\{x_n\} = \frac{3z}{z^2 - 9}.$$

(a)
$$\begin{aligned}
Z\{te^{3t}\} &= Te^{3T}z^{-1} + 2Te^{6T}z^{-2} + 3Te^{9T}z^{-3} + \ldots \\
&= Te^{3T}z^{-1}(1 + 2e^{3T}z^{-1} + 3e^{6T}z^{-2} + \ldots) \\
&= Te^{3T}z^{-1}(1 - e^{3T}z^{-1})^{-2} \\
&= \frac{Te^{3T}z}{(z - e^{3T})^2}.
\end{aligned}$$

The above result can be confirmed by using Table 2.

$$Z\{t\} = \frac{Tz}{(z-1)^2}.$$

Therefore $Z\{e^{3T}t\} = \dfrac{Te^{-3T}z}{(e^{-3T}z - 1)^2} = \dfrac{Te^{3T}z}{(z - e^{3T})^2}.$

(b) Expressing the function as the sum of partial functions *after* taking out z as a factor we obtain

$$\begin{aligned}
\frac{3z}{z^2 - 9} &= \frac{z}{2}\left(\frac{1}{z-3} - \frac{1}{z+3}\right) \\
&= \frac{1}{2}\left(\frac{z}{z-3} - \frac{z}{z+3}\right).
\end{aligned}$$

Using Table 2 we obtain

$$x(t) = \frac{1}{2}(3^{t/T} - (-3)^{t/T}), \quad t = 0, T, 2T, 3T, \ldots.$$

But $t = nT$. Therefore the sequence $\{x_n\}$ is given by

$$x_n = \frac{1}{2}(3^n - (-3)^n), \quad n = 0, 1, 2, 3, \ldots$$

Example 5.5
Prove property (iii) of 'further properties ...' above.

Property (iii) is

$$Z\{f(t+T)\} = zZ\{f(t)\} - zf(0).$$

$$Z\{f(t+T)\} = f(T) + f(2T)z^{-1} + f(3T)z^{-2} + \ldots$$

$$= z(f(0) + f(T)z^{-1} + f(2T)z^{-2} + f(3T)z^{-3} + \ldots) - zf(0)$$

$$= zZ\{f(t)\} - zf(0).$$

Problems

(1) Find the sampled sequence $f(t)$, $t = 0, T, 2T, 3T, \ldots$ such that

(i) $\quad Z\{f(t)\} = \dfrac{3z^3 - 8z^2 + 9z}{(z-2)(z+3)(z-1)}$

(ii) $\quad Z\{f(t)\} = \dfrac{3z^3 - 6z^2 - 5z}{(z-1)^2(z+1)}.$

For (i) and (ii) find the sequence $\{x_n\}$ which has the same z-transform.

(2) Use the definition of a z-transform to prove that

$$Z\{(t-T)u(t-T)\} = \dfrac{T}{(z-1)^2}.$$

The above result is a special case of property (ii). Prove this property.

(3) Prove properties (iv) and (v).

(4) Use Table 2 to write down the z-transform of

$$e^{jbt}, \; t = 0, T, 2T, \ldots.$$

Hence obtain the z-transforms of $\cos bt$ and $\sin bt$ in the forms given in Table 2.

(5) Prove that

$$Z\{e^{ct}f(t)\}' = F(e^{-cT}z).$$

Hence find the z-transforms of $e^{ct}\cos bt$ and $e^{ct}\sin bt$.

(6) Prove properties (vii) and (viii).

Answers to problems

(1) (i) $\;f(t) = 2^{t/T} + 3(-3)^{t/T} - 1, \; t = 0, T, 2T, 3T, \ldots,$

$$x_n = 2^n + 3(-3)^n - 1, \; n = 0, 1, 2, 3, \ldots.$$

(ii) $\;f(t) = 2 - 4\left(\dfrac{t}{T}\right) + (-1)^{t/T}, \; t = 0, T, 2T, 3T, \ldots,$

$$x_n = 2 - 4n + (-1)^n, \; n = 0, 1, 2, 3, \ldots.$$

(5) $\;Z\{e^{ct}\cos bt\} = \dfrac{z^2 - ze^{cT}\cos bT}{z^2 - 2ze^{cT}\cos bT + e^{2cT}}$

$$Z\{e^{ct}\sin bt\} = \dfrac{ze^{cT}\sin bT}{z^2 - 2ze^{cT}\cos bT + e^{2cT}}.$$

CHAPTER 6

Solution of linear difference equations

6.1 INTRODUCTION

The concept of a digital system was introduced in section 5.2. When such a system is *linear* the output sequence $\{y_n\}$ and input sequence $\{x_n\}$ may be related by a linear difference equation.

In this chapter z-transform techniques are applied to the solution of linear difference equations. For simple cases z-transforms are not necessarily the best method of solution, but the corresponding transfer function notation is useful in subsequent work. Attention will be drawn to the similarities between the transfer function approaches to differential equations and difference equations. In particular the frequency characteristics of discrete and continuous systems will be compared.

6.2 DIFFERENCE EQUATIONS

A difference equation expresses a particular output y_n of a digital system in terms of the synchronous input x_n and earlier outputs and inputs, for example

$$y_n = \sum_{i=1}^{N} a_i y_{n-i} + \sum_{i=0}^{M} b_i x_{n-i}, \ a_N \neq 0, \ b_M \neq 0. \tag{6.1}$$

The integer N is the **order** of the difference equation. It is assumed that for $n < 0$, $x_n = y_n = 0$.

Using the delay property $Z\{x_{n-k}\} = z^{-k} Z\{x_n\}$ (property (iv) in Section 5.5), we can transform Equation (6.1) to obtain

$$Z\{y_n\} = \sum_{i=1}^{N} a_i z^{-i} Z\{y_n\} + \sum_{i=0}^{M} b_i z^{-i} Z\{x_n\},$$

that is, $\quad Z\{y_n\} = \left(\sum_{i=0}^{M} b_i z^{-i} \middle/ \left(1 - \sum_{i=1}^{N} a_i z^{-i}\right) \right) Z\{x_n\},$

that is, $\quad Z\{y_n\} = H(z) \cdot Z\{x_n\},$

where $\quad H(z) = \sum_{i=0}^{M} b_i z^{N-i} \middle/ \left(z^N - \sum_{i=1}^{N} a_i z^{N-i} \right)$

is called the **transfer function** of the digital system.

If all the constants a_i, $i = 1, 2, 3, \ldots, N$ are zero then the system is said to be **non-recursive.** Otherwise the system is **recursive.** The worked examples show the use of z-transforms in the solution of difference equations.

Three particular input sequences are worth mentioning.

(i)   The **unit step** sequence $u_n = 1$, $n = 0, 1, 2, 3, \ldots$,

$$Z\{u_n\} = \frac{z}{z-1}.$$

(ii)   The **unit impulse** sequence $\delta_n = \begin{cases} 1, \ n = 0 \\ 0, \ n = 1, 2, 3, \ldots, \end{cases}$

$$Z\{\delta_n\} = 1.$$

(iii)   The **delayed unit impulse** $\delta_{n-k} = \begin{cases} 0, \ n = 1, 2, 3, \ldots k-1 \\ 1, \ n = k \\ 0, \ n = k+1, k+2, \ldots \end{cases}$

$$Z\{\delta_{n-k}\} = z^{-k},$$

Note that the output of a digital system with transfer function $H(z)$ for a unit impulse input is given by

$$Z\{y_n\} = H(z) \cdot 1,$$

that is $y_n = h_n$ where $\{h_n\} = Z^{-1}\{H(z)\}$.

The sequence $\{h_n\}$ is called the **impulse response sequence.**

When the input is $\{\delta_{n-k}\}$ the digital system will have output

$$\{y_n\} = Z^{-1}\{z^{-k}H(z)\} = \{h_{n-k}\},$$

(where $h_i = 0$ for $i < 0$).

Further, since any general input sequence $\{x_n\}$ can be expressed as the infinite sum

$$\{x_n\} = x_0\{\delta_n\} + x_1\{\delta_{n-1}\} + x_2\{\delta_{n-2}\} + \ldots,$$

then the output of the system $H(z)$ for input $\{x_n\}$ is given by

$$\{y_n\} = x_0\{h_n\} + x_1\{h_{n-1}\} + x_2\{h_{n-2}\} + \ldots,$$

that is,
$$y_n = \sum_{i=0}^{n} x_i h_{n-i}, \ n = 0, 1, 2, 3, \ldots.$$

This is the *convolution property* for digital systems which closely resembles the corresponding property for continuous systems given in section 1.6.

Example 6.1

Solve the difference equation

$$y_{n+2} + 3y_{n+1} + 2y_n = 0, \quad n = 0, 1, 2, 3, \ldots,$$

given that $y_0 = 1$, $y_1 = 2$.

We can of course obtain the numerical values $y_2, y_3, y_4, \ldots$, simply by substitution in the given equation.

$y_2 + 3y_1 + 2y_0 = 0$, therefore $y_2 = -6 - 2 = -8$
$y_3 + 3y_2 + 2y_1 = 0$, therefore $y_3 = 24 - 4 = 20$
$y_4 + 3y_3 + 2y_2 = 0$, therefore $y_4 = -60 + 16 = -44$
etc.

However, z-transform techniques allow us to find a general formula for y_n. Using property (v) in section 5.5,

$$Z\{y_{n+2}\} = z^2 Z\{y_n\} - (z^2 y_0 + z y_1)$$

$$= z^2 Z\{y_n\} - z^2 - 2z,$$

$$Z\{y_{n+1}\} = z Z\{y_n\} - z y_0$$

$$= z Z\{y_n\} - z.$$

Transforming the given difference equation gives

$$z^2 Z\{y_n\} - z^2 - 2z + 3(z Z\{y_n\} - z) + 2Z\{y_n\} = 0.$$

Therefore $Z\{y_n\} = \dfrac{z^2 + 5z}{z^2 + 3z + 2} = \dfrac{z(z + 5)}{(z + 1)(z + 2)}.$

Applying partial fractions but retaining the factor z we obtain

$$Z\{y_n\} = z\left(\frac{4}{z + 1} - \frac{3}{z + 2}\right)$$

$$= \frac{4z}{z - (-1)} - \frac{3z}{z - (-2)}.$$

From Table 2,

$$y_n = 4(-1)^n - 3(-2)^n, \quad n = 0, 1, 2, 3, \ldots.$$

Example 6.2

Solve the difference equation

$$y_{n+2} + 3y_{n+1} + 2y_n = x_n, \quad n = 0, 1, 2, 3, \ldots,$$

where $x_n = n$ and $y_0 = A$, $y_1 = B$.

Transforming the difference equation,

$$z^2 Z\{y_n\} - z^2 A - zB + 3(Z\{y_n\} - zA) + 2Z\{y_n\} = \frac{z}{(z-1)^2}.$$

Therefore $Z\{y_n\} = \dfrac{z}{(z-1)^2(z+1)(z+2)} + \dfrac{z(Az + B + 3A)}{(z+1)(z+2)}$

$$= \frac{1}{6} \cdot \frac{z}{(z-1)^2} - \frac{5}{36} \cdot \frac{z}{(z-1)} + \frac{C_1 z}{(z+1)} + \frac{C_2 z}{(z+2)}$$

where $C_1 = 2A + B + \dfrac{1}{4}$, $C_2 = -(A+B) - \dfrac{1}{9}$.

Again from Table 2,

$$y_n = \frac{n}{6} - \frac{5}{36} + C_1(-1)^n + C_2(-2)^n, \; n = 0, 1, 2, 3, \ldots$$

(Compare this answer with Example 6.1. Note the 'particular solution', $\dfrac{n}{6} - \dfrac{5}{36}$, and the 'complementary function', $C_1(-1)^n + C_2(-2)^n$. The latter contains the numbers -1 and -2 which are roots of the 'characteristic equation', $z^2 + 3z + 2 = 0$).

Example 6.3

A digital system is described by the transfer function $\dfrac{z^2 + 3z}{z^2 + 4z - 21}$. Find the output sequence for a unit step input sequence.

———

Fig. 6.1 shows the system in block diagram notation.

$$Z\{u_n\} = \frac{z}{(z-1)}.$$

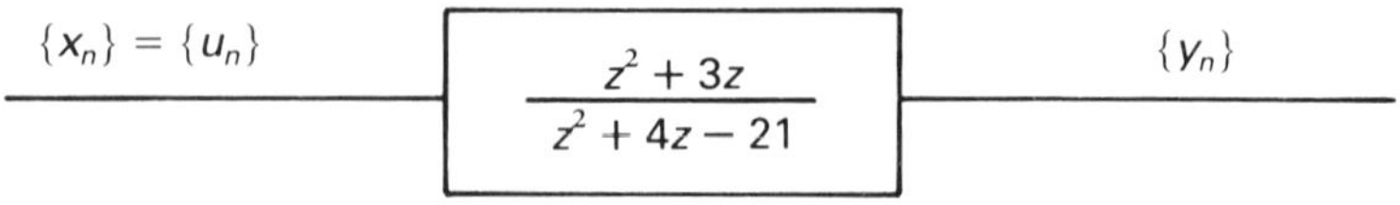

$\{x_n\} = \{u_n\}$ $\dfrac{z^2 + 3z}{z^2 + 4z - 21}$ $\{y_n\}$

Fig. 6.1

Therefore $Z\{y_n\} = \left(\dfrac{z^2 + 3z}{z^2 + 4z - 21}\right) \cdot \left(\dfrac{z}{z - 1}\right)$

$$= \frac{z(z^2 + 3z)}{(z - 1)(z + 7)(z - 3)}$$

$$= -\frac{1}{4} \cdot \frac{z}{(z - 1)} + \frac{7}{20} \cdot \frac{z}{(z + 7)} + \frac{9}{10} \cdot \frac{z}{(z - 3)}.$$

Therefore $y_n = -\dfrac{1}{4} + \dfrac{7}{20}(-7)^n + \dfrac{9}{10} \cdot 3^n$, $n = 0, 1, 2, \ldots$.

Alternatively we can obtain the first few numerical values by writing down the corresponding difference equation as follows.

$$\frac{Z\{y_n\}}{Z\{x_n\}} = \frac{z^2 + 3z}{z^2 + 4z - 21} = \frac{1 + 3z^{-1}}{1 + 4z^{-1} - 21z^{-2}}.$$

Therefore $y_n + 4y_{n-1} - 21y_{n-2} = x_n + 3x_{n-1}$

and $y_n = x_n + 3x_{n-1} - 4y_{n-1} + 21y_{n-2}$, (recursive system).

Taking $y_{-1} = y_{-2} = x_{-1} = 0$,

$$\begin{aligned}
y_0 &= x_0 = 1, \\
y_1 &= x_1 + 3x_0 - 4y_0 = 1 + 3 - 4 = 0, \\
y_2 &= x_2 + 3x_1 - 4y_1 + 21y_0 \\
&= 1 + 3 - 0 + 21 = 25,
\end{aligned}$$

etc.

Example 6.4

Given that the function $f(t) = e^{-at} \sin bt$ is sampled at times $t = 0, T, 2T, 3T, \ldots$, confirm that

$$Z\{f(t)\} = \frac{z\,e^{-aT} \sin bT}{z^2 - 2z\,e^{-aT} \cos bT + e^{-2aT}}.$$

Hence solve the difference equation

$$y(t + 2T) - y(t + T) + 0.5y(t) = 1$$

for $t = 0, T, 2T, 3T, \ldots$,

given that $y(0) = 2$, $y(T) = 3$.

From Table 2

$$Z\{\sin bt\} = \frac{z \sin bT}{z^2 - 2z \cos bT + 1}.$$

Using property (vi) in section 5.5,

$$Z\{e^{-at}\sin bt\} = \frac{z\,e^{aT}\sin bT}{z^2\,e^{2aT} - 2z\,e^{aT}\cos bT + 1}$$

$$= \frac{z\,e^{-aT}\sin bT}{z^2 - 2z\,e^{-aT}\cos bT + e^{-2aT}}$$

Transforming the difference equation,

$$z^2 Z\{y(t)\} - z^2 2 - z3 - (zZ\{y(t)\} - z2) + 0.5Z\{y(t)\} = \frac{z}{z-1}.$$

Therefore $(z^2 - z + 0.5)Z\{y(t)\} = 2z^2 + z + \dfrac{z}{z-1} = \dfrac{z(2z^2 - z)}{(z-1)}.$

Therefore $Z\{y(t)\} = \dfrac{z(2z^2 - z)}{(z-1)(z^2 - z + 0.5)}$

$$= \frac{2z}{(z-1)} + \frac{z}{z^2 - z + 0.5}.$$

Comparing the second term with the transform

$$Z\{e^{-at}\sin bt\} = \frac{z\,e^{-aT}\sin bT}{z^2 - 2z\,e^{-aT}\cos bT + e^{-2aT}},$$

we have $e^{-2aT} = 0.5$, that is, $e^{-aT} = \dfrac{1}{\sqrt{2}}.$

Also $2e^{-aT}\cos bT = 1$, therefore $\cos bT = \dfrac{1}{\sqrt{2}},$

that is, $bT = \dfrac{\pi}{4}$, $b = \dfrac{\pi}{4T}$, and $\sin bT = \dfrac{1}{\sqrt{2}}.$

Further $e^{-at} = (e^{-aT})^{t/T} = \left(\dfrac{1}{\sqrt{2}}\right)^{t/T}.$

It follows that

$$Z\left\{\left(\frac{1}{\sqrt{2}}\right)^{t/T}\sin\left(\frac{\pi t}{4T}\right)\right\} = \frac{z/2}{z^2 - z + 0.5}.$$

Therefore $y(t) = 2 + 2\left(\dfrac{1}{\sqrt{2}}\right)^{t/T}\sin\left(\dfrac{\pi t}{4T}\right)$, $t = 0, T, 2T, 3T, \ldots$

or $y_n = 2 + 2\left(\dfrac{1}{\sqrt{2}}\right)^{n}\sin\dfrac{n\pi}{4}$, $n = 0, 1, 2, 3, \ldots$

Problems

(1) Solve the difference equation

$$y_{n+2} - 4y_n = 0, \ y_0 = 1, \ y_1 = -1.$$

(2) Show that $Z\{na^n\} = \dfrac{az}{(z-a)^2}.$

Solve the difference equation

$$f_{n+2} + 2f_{n+1} + f_n = 0, \ f_0 = 1, \ f_1 = 0.$$

(3) The function $x(t)$ is defined for $t = 0, 1, 2, 3, \ldots$.
Obtain the solution of

$$x(t+1) - x(t) = 2t + 1, \ x(0) = C.$$

(4) A digital system is described by the transfer function $\dfrac{z(z+1)}{(z+2)^2}.$

Obtain the output sequence for
(i) a unit impulse squence,
(ii) a unit step sequence.
Check your answers for $n = 0, 1, 2, 3$ by writing down the corresponding difference equation.

(5) Obtain the output sequence of the system shown in Fig. 6.2 when the sampler operates at $t = 0, 0.1, 0.2, 0.3, \ldots$.

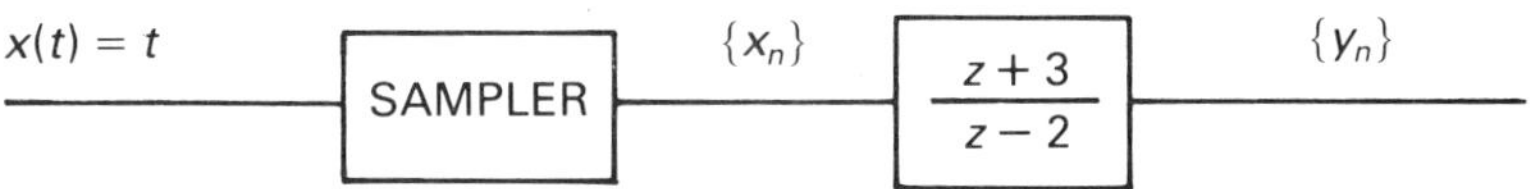

Fig. 6.2

(6) Show that

$$Z^{-1}\left\{\frac{z+3}{z-2}\right\} = \begin{cases} 1, & n = 0 \\ 5(2)^{n-1}, & n = 1, 2, 3, \ldots \end{cases}$$

Use the convolution property to confirm the result of problem 5 for $n = 0, 1, 2, 3, 4, 5$.

(7) Show that

(i) $Z\{e^{-at}\cos bt\} = \dfrac{z^2 - ze^{-aT}\cos bT}{z^2 - 2ze^{-aT}\cos bT + e^{-2aT}},$

(ii) $Z^{-1}\left\{\dfrac{z^2}{z^2 - 2z + 2}\right\} = (\sqrt{2})^{t/T}\left(\cos\dfrac{\pi t}{4T} + \sin\dfrac{\pi t}{4T}\right),$

$$t = 0, T, 2T, 3T, \ldots$$

(8) Solve the difference equation

$$x(t+2T)-x(t+T)+0.5x(t) = 1, \quad t = 0, T, 2T, 3T, \ldots,$$

given that $x(0) = x(T) = 1$.

(9) Solve the difference equation

$$x(t+2)+2x(t+1)+2x(t) = 5t, \quad t = 0, 1, 2, 3, \ldots,$$

given that $x(0) = x(1) = 0$.

Answers to problems

(1) $y_n = \frac{1}{4}(2^n + 3(-2)^n)$.

(2) $f_n = (-1)^n(1-n)$.

(3) $x(t) = t^2 + C, \quad t = 0, 1, 2, 3, \ldots.$

(4) (i) $(-2)^n\left(1+\dfrac{n}{2}\right), \quad n = 0, 1, 2, 3, \ldots.$

(ii) $\dfrac{2}{9} + \dfrac{(-2)^n}{9}(7+3n), \quad n = 0, 1, 2, 3, \ldots.$

(5) $y_n^* = -0.5 - 0.4\,n + 0.5(2)^n$.

(8) $x(t) = 2 - \left(\dfrac{1}{\sqrt{2}}\right)^{t/T}\left(\cos\dfrac{\pi t}{4T} + \sin\dfrac{\pi t}{4T}\right), \quad t = 0, T, 2T, 3T, \ldots.$

(9) $x(t) = t - \dfrac{4}{5} + (\sqrt{2})^t\left(\dfrac{4}{5}\cos\dfrac{3\pi t}{4} + \dfrac{3}{5}\sin\dfrac{3\pi t}{4}\right), \quad t = 0, 1, 2, 3 \ldots.$

6.3 COMPARISON BETWEEN CONTINUOUS AND DISCRETE SYSTEMS

In Chapters 2 and 3 we saw how important results, such as stability and steady-state solutions, could be obtained for continuous systems by consideration of the system transfer function without the necessity to obtain a complete solution of the corresponding differential equation. Similar techniques may be applied to digital systems once the system transfer function $H(z)$ has been established, and information about system stability and steady solution can be obtained without solving the system difference equation.

The techniques depend on the relationship defined earlier between the parameters s and z

$$z = e^{sT} \tag{6.2}$$

where the input sequence $\{x_n\}$ is considered to have been obtained by sampling some function $x(t)$ at times $t = 0, T, 2T, \ldots,$

that is, $\{x_n\} = \{x(nT)\}$.

Consider, for example, the digital system with transfer function

$$H(z) = \frac{3z^2 + 2z}{2z^2 + 5z - 3}.$$

The output of the system will be obtained from

$$Z\{y_n\} = \left(\frac{3z^2 + 2z}{(2z - 1)(z + 3)} \right) Z\{x_n\}$$

where $\{x_n\}$ is the input sequence. Clearly $Z\{y_n\}$ will contain terms depending on the input (the 'steady' solution) and terms $\dfrac{z}{z - 0.5}, \dfrac{z}{z + 3}$ arising from the system. These latter terms cause the output sequence to include $(0.5)^n$ and $(-3)^n$. Now $(0.5)^n \to 0$ as $n \to \infty$, but $(-3)^n$ is unbounded as $n \to \infty$. This particular system is said to be *unstable* because the part of the output sequence arising from the system itself (the 'transient' solution) does not tend to a finite limit as $n \to \infty$.

In general, if $H(z)$ contains a *pole* at $z = a$, then $Z\{y_n\}$ will contain a term $\dfrac{z}{z - a}$ and y_n will include a^n. For stability *all* poles of $H(z)$ must be such that $|z| < 1$, (that is, $a^n \to 0$ as $n \to \infty$).

The *stability region* of the complex z-plane is shown in Fig. 6.3 together with the stability region of the complex s-plane applicable to continuous systems (see section 2.6). The relationship between the two regions arises directly from equation (6.2).

Writing $s = \sigma + j\omega$, the continuous system stability region is $\sigma < 0$.

Now $z = x + jy = e^{sT} = e^{\sigma T}(\cos \omega T + j \sin \omega T)$, and $|z|^2 = x^2 + y^2 = e^{2\sigma T}$. Therefore $\sigma < 0$ corresponds to $|z| < 1$, the digital system stability region.

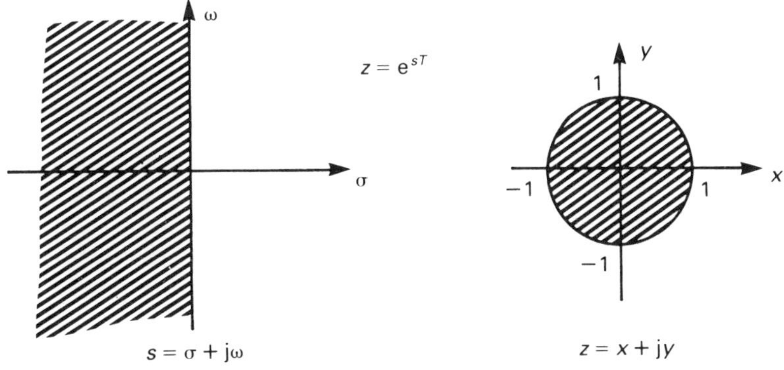

Fig. 6.3 Stability regions (shaded) for s-plane and z-plane.

Using the notation of section 2.6 the output sequence of a digital system is given by

$$\{y_n\} = \{y_n\}_{TR} + \{y_n\}_{SS}$$

where $\{y_n\}_{TR}$ is the *transient response* of the system and contains terms independent of the input sequence $\{x_n\}$. For a *stable* system $\{y_n\}_{TR} \to 0$ as $n \to \infty$. $\{y_n\}_{SS}$ is the *steady-state* solution which depends on the input sequence, and for a stable system $\{y_n\} \to \{y_n\}_{SS}$ as $n \to \infty$. For example, consider the system $H(z) = \dfrac{z+1}{(z-\frac{1}{2})(z+\frac{1}{3})}$ with input $\{x_n\} = \{u_n\}$, the unit step sequence. The output sequence is given by

$$Z\{y_n\} = \frac{z(z+1)}{(z-1)(z-\frac{1}{2})(z+\frac{1}{3})}$$

$$= \frac{3z}{(z-1)} - \frac{18z}{5(z-\frac{1}{2})} + \frac{3z}{5(z+\frac{1}{3})}$$

Therefore $y_n = 3 - \dfrac{18}{5}\left(\dfrac{1}{2}\right)^n + \dfrac{3}{5}\left(-\dfrac{1}{3}\right)^n$, $n = 0, 1, 2, 3, \ldots$.

Here $\{y_n\}_{TR} = \left\{ -\dfrac{18}{5}\left(\dfrac{1}{2}\right)^n + \dfrac{3}{5}\left(-\dfrac{1}{3}\right)^n \right\}$

and $\{y_n\}_{SS} = 3\{u_n\}$, (that is, $y_n \to 3$ as $n \to \infty$).

In section 2.6 we noted that y_{SS} for a continuous system with unit step function input was obtained immediately by putting $s = 0$ in the system transfer function. Similarly, for a digital system $H(z)$ the steady value of the output for a unit step input sequence is obtained by putting $z = 1$ in $H(z)$, since from equation (6.2) $z = e^{sT} = 1$ when $s = 0$. Therefore in this case $y_n \to H(1)$ as $n \to \infty$, that is, $\{y_n\}_{SS} = H(1)\{u_n\}$. (This result can also be derived from the final value theorem for z-transforms, property (ix) in section 5.5.)

In the above example $H(z) = \dfrac{(z+1)}{(z-\frac{1}{2})(z+\frac{1}{3})}$ and, for

$$\{x_n\} = \{u_n\}, \; y_n \to \frac{2}{\frac{1}{2}\left(\frac{4}{3}\right)} = 3 \text{ as } n \to \infty.$$

A similar short cut applies when the input sequence is a sampled sine (or cosine) wave. In Chapter 3 for continuous systems we used the substitution $s = j\omega$. For digital systems we set $z = e^{j\omega T}$ in the system transfer function $H(z)$. For the sampled input $\{x_n\} = \{a \sin \omega nT\}$ the steady-state output sequence is given by $\{y_n\}_{SS} = \{aR \sin(\omega nT + \phi)\}$ where R and ϕ are defined by $H(e^{j\omega T}) = Re^{j\phi}$.

Summarizing the results of this section applied to the system shown in Fig. 6.4;

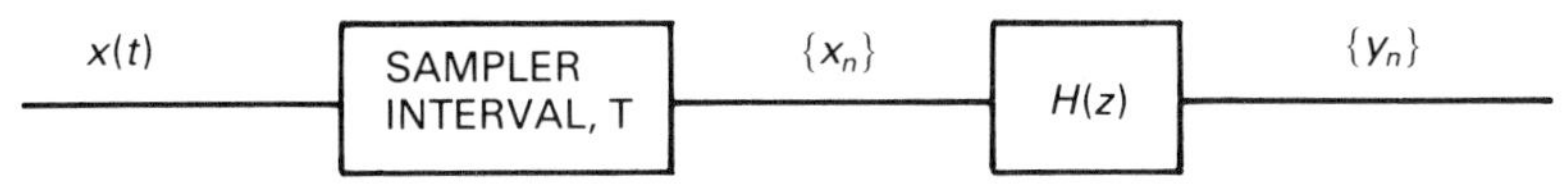

Fig. 6.4

(i)   The system is stable if all poles of $H(z)$ lie *inside* the unit circle $|z| = 1$.

(ii) For $\{x_n\} = K\{u_n\}$, $\{y_n\}_{SS} = KH(1)\{u_n\}$.

(iii) For $x(t) = a \sin \omega t$, $\{y_n\}_{SS} = \{aR \sin(\omega nT + \phi)\}$, where $H(e^{j\omega T}) = Re^{j\phi}$.

All of these properties are related to the corresponding continuous system properties by application of equation (6.2).

Example 6.5

Suppose that $T = 1$ and $H(z) = \dfrac{z}{z - \frac{1}{2}}$ in Fig. 6.4. Write down the steady-state output sequence for the cases

(i) $\{x_n\} = 4\{u_n\}$,

(ii) $x(t) = \sin \dfrac{\pi}{3} t$.

(i)   The input sequence is $4\{u_n\}$.

Therefore $\{y_n\}_{SS} = 4H(1)\{u_n\}$

$$= 8\{u_n\}.$$

(ii) The input sequence is $\left\{ \sin \dfrac{\pi}{3} n \right\}$,

that is, $T = 1$, $\omega = \dfrac{\pi}{3}$ and we set $z = \exp(j\pi/3)$.

Therefore $H(\exp(j\pi/3)) = \dfrac{\exp(j\pi/3)}{\exp(j\pi/3) - \frac{1}{2}}$

$$= \frac{\exp(j\pi/3)}{j\dfrac{\sqrt{3}}{2}}$$

$$= \frac{2\exp(j\pi/3)}{\sqrt{3}\exp(j\pi/2)} = \frac{2}{\sqrt{3}} \exp(-j\pi/6).$$

It follows that $R = \dfrac{2}{\sqrt{3}}$, $\phi = -\dfrac{\pi}{6}$

and $\{y_n\} = \left\{\dfrac{2}{\sqrt{3}} \sin\left(\dfrac{\pi}{3}n - \dfrac{\pi}{6}\right)\right\}$.

Example 6.6

Given the difference equation

$$y_n + 2y_{n-1} + 2y_{n-2} = x_{n-1} + x_{n-2}, \quad x_n = y_n = 0 \text{ for } n < 0,$$

obtain the steady-state solution for the case

$$\{x_n\} = \{2,\, 0,\, -2,\, 0,\, 2,\, 0,\, -2,\, 0,\, \ldots\}.$$

The difference equation can be written in the form

$$(1 + 2z^{-1} + 2z^{-2})Z\{y_n\} = (z^{-1} + z^{-2})Z\{x_n\},$$

that is, $H(z) = \dfrac{Z\{y_n\}}{Z\{x_n\}} = \dfrac{z+1}{z^2 + 2z + 2}$

is the corresponding system transfer function.

The input sequence can be considered as the function $x(t) = 2\cos t$ sampled at

$$t = 0,\, \frac{\pi}{2},\, \pi,\, \frac{3\pi}{2},\, 2\pi,\, \ldots,\, \text{ that is, } T = \frac{\pi}{2},\, \omega = 1.$$

Therefore $H(e^{j\omega T}) = H(j) = \dfrac{1+j}{1+2j} = Re^{j\phi}$.

where $R = \sqrt{\dfrac{2}{5}}$, $\phi = \dfrac{\pi}{4} - \tan^{-1} 2$.

It follows that

$$\{y_n\}_{ss} = \left\{2\sqrt{\frac{2}{5}}\cos\left(n\frac{\pi}{2} + \frac{\pi}{4} - \tan^{-1} 2\right)\right\}.$$

Problems

(1) For the system shown in Fig. 6.4 suppose that

$$T = \frac{\pi}{2} \text{ and } H(z) = \frac{z}{4z^2 - 1}.$$

Confirm that the system is stable, and obtain the steady-state output sequence for the cases
(i) $x(t) = u(t)$, the unit step function,
(ii) $x(t) = 5 \sin t$.

(2) Which of the following systems are stable?

(i) $H(z) = \dfrac{z^2 + 3z + 2}{z(6z^2 + 5z + 1)}$,

(ii) $H(z) = \dfrac{z + 1}{z^4 - 0.25}$,

(iii) $H(z) = \dfrac{3z^2 + 2z}{4z^4 + 8z^3 + 9z^2 + 2z + 2}$.

For the stable systems, calculate the steady-state output sequence when the input sequence is the unit step sequence.

(3) Obtain the steady-state solution of the difference equation

$$y_n = \frac{1}{5}\left(2 \sin \frac{n\pi}{2} + 3y_{n-1}\right), \ n = 0, 1, 2, 3, \ldots,$$

where $y_{-1} = 0$.

(4) Write down the steady-state output sequence for the digital system satisfying the difference equation.

$$2y_n = x_{n-3} + y_{n-1} - 0.25y_{n-3} + 0.125y_{n-4}, \ n = 0, 1, 2, 3, \ldots,$$

(where x_n and y_n are 0 for $n < 0$), given that

$$\{x_n\} = \{1, \ -1, \ 1, \ -1, \ 1, \ -1, \ldots\}.$$

Answers to problems

(1) (i) $\dfrac{1}{3}\{u_n\}$ (ii) $\left\{\sin(n-1)\dfrac{\pi}{2}\right\}$

(2) (i) and (ii) are stable.

(i) $\dfrac{1}{2}\{u_n\}$ (ii) $\dfrac{8}{3}\{u_n\}$

(3) $\dfrac{2}{\sqrt{34}}\sin\left(n\dfrac{\pi}{2} - \tan^{-1}0.6\right)$

(4) $\left\{-\dfrac{8}{21}, \dfrac{8}{21}, \ -\dfrac{8}{21}, \dfrac{8}{21}, \ldots\right\}$.

6.4 ALTERNATIVE METHODS FOR INVERSE z-TRANSFORMS

In the previous sections, z-transforms were inverted using Table 2 after conversion to partial fraction form. Alternative methods shown in the following worked examples are:

(i) power series inversion,
(ii) long division.

Example 6.7

Given that $F(z) = \dfrac{z+3}{z-2}$ find the sequence $\{f_n\}$ by

(i)   using Table 2,
(ii) power series inversion,
(iii) long division.

(i) $F(z) = \dfrac{z}{z-2} + \dfrac{3}{z-2}$

    Therefore $f_n = 2^n + 3(2)^{n-1}\, u(n-1)$,

    that is, $f_0 = 1; f_n = 5(2)^{n-1}$, $n = 1, 2, 3, \ldots$.

(ii) $F(z) = \dfrac{1 + 3z^{-1}}{1 - 2z^{-1}}$

$$= (1 + 3z^{-1})(1 + 2z^{-1} + 2^2 z^{-2} + 2^3 z^{-3} + \ldots)$$
$$= 1 + (3 + 2)\, z^{-1} + (3.2 + 2^2)\, z^{-2} + (3.2^2 + 2^3)\, z^{-3} + \ldots$$

$$= 1 + \sum_{n=1}^{\infty} (3(2)^{n-1} + 2^n)\, z^{-n}.$$

    Therefore it follows from the definition of a z-transform that

$$f_0 = 1; f_n = 5(2)^{n-1}, \; n = 1, 2, 3, \ldots.$$

(iii)
$$\begin{array}{r}
1 + 5z^{-1} + 10z^{-2} + 20z^{-3} + \ldots \\[4pt]
z-2 \, \overline{\smash{)}\, z + 3 \phantom{+ 3 + 20z^{-3} + \ldots}} \\
\underline{z - 2} \phantom{+ 3 + 20z^{-3} + \ldots} \\
5 \phantom{+ 20z^{-3} + \ldots} \\
5 - 10z^{-1} \\
\underline{} \\
10z^{-1} \\
10z^{-1} - 20z^{-2} \\
\underline{\phantom{10z^{-1} - 2}} \\
20z^{-2} \text{ etc.}
\end{array}$$

Therefore $f_0 = 1$, $f_1 = 5$, $f_2 = 10$, $f_3 = 20$, $\ldots$.

This method is useful when only the first few numbers in the sequence are required, although in this example it is not difficult to obtain the general term.

Example 6.8

Given $F(z) = \dfrac{z^3 - 9z^2 + 5z - 1}{4z^3 - 8z^2 + 5z - 1}$,

find f_n for $n = 0, 1, 2$, using (i) power series inversion, (ii) long division.

(i) $F(z) = \dfrac{z^3 - 9z^2 + 5z - 1}{(z-1)(2z-1)^2} = \dfrac{1 - 9z^{-1} + 5z^{-2} - z^{-3}}{4(1 - z^{-1})(1 - 2^{-1}z^{-1})^2}$

$$= \frac{1}{4}(1 - 9z^{-1} + 5z^{-2} - z^{-3})(1 + z^{-1} + z^{-2} + \ldots)$$

$$\times \left(1 + \frac{2}{2}z^{-1} + \frac{3}{2^2}z^{-2} + \ldots\right)$$

$$= \frac{1}{4}\left(1 + \left(1 + \frac{2}{2} - 9\right)z^{-1} + \left(1 + \frac{3}{2^2} + 5 - 9 - 9 + \frac{2}{2}\right)z^{-2} + \ldots\right)$$

Therefore $f_0 = \dfrac{1}{4}$, $f_1 = -\dfrac{7}{4}$, $f = -\dfrac{41}{16}$.

(ii) $4z^3 - 8z^2 + 5z - 1 \overline{\smash{\big)}\ z^3 - 9z^2 + 5z - 1}$ $\frac{1}{4} - \frac{7}{4}z^{-1} - \frac{41}{16}z^{-2} + \ldots$

$$z^3 - 2z^2 + \tfrac{5}{4}z - \tfrac{1}{4}$$

$$-7z^2 + \tfrac{15}{4}z - \tfrac{3}{4}$$

$$-7z^2 + 14z - \tfrac{35}{4} + \tfrac{7}{4}z^{-1}$$

$$-\tfrac{41}{4}z + 8 - \tfrac{7}{4}z^{-1} \text{ etc.}$$

Therefore $f_0 = \dfrac{1}{4}$, $f_1 = -\dfrac{7}{4}$, $f = -\dfrac{41}{16}$.

For this example it is not easy to get a general form for f_n without using a partial fraction method.

Example 6.9

The input $\{x_n\}$ to a digital system with transfer function $\dfrac{z+1}{2z-1}$ is given by $x_n = n$. Obtain the output sequence $\{y_n\}$ using power series inversion.

$$Z\{x_n\} = \frac{z}{(z-1)^2}.$$

Therefore $Z\{y_n\} = \dfrac{z(z+1)}{(z-1)^2\,(2z-1)}$

$$= \frac{z^{-1}+z^{-2}}{2(1-z^{-1})^2\,(1-2^{-1}z^{-1})}$$

$$= \frac{1}{2}(z^{-1}+z^{-2})\,(1+2z^{-1}+3z^{-2}+4z^{-3}+\ldots)$$

$$\times\left(1+\frac{z^{-1}}{2}+\frac{z^{-2}}{2^2}+\frac{z^{-3}}{2^3}+\ldots\right)$$

$$= \frac{1}{2}(z^{-1}+z^{-2})\left(1+\left(2+\frac{1}{2}\right)z^{-1}+\left(3+\frac{2}{2}+\frac{1}{2^2}\right)z^{-2}\right.$$

$$\left.+\left(4+\frac{3}{2}+\frac{2}{2^2}+\frac{1}{2^3}\right)z^{-3}+\ldots\right)$$

Therefore $y_0 = 0$

$$y_1 = \frac{1}{2}$$

$$y_2 = \frac{1}{2}\left(1+2+\frac{1}{2}\right)$$

$$y_3 = \frac{1}{2}\left(2+\frac{1}{2}+3+\frac{2}{2}+\frac{1}{2^2}\right)$$

$$y_4 = \frac{1}{2}\left(3+\frac{2}{2}+\frac{1}{2^2}+4+\frac{3}{2}+\frac{2}{2^2}+\frac{1}{2^3}\right)$$

that is, $$y_n = \frac{1}{2}\sum_{i=1}^{n}\frac{2i-1}{2^{n-i}}.$$

(This general form is not as convenient as the form obtained by the partial fraction approach, $y = 2n-3+3\,(\tfrac{1}{2})^n$. In general, the power series approach (and the long division approach) is useful when the first few numbers are required, and sometimes leads to a formula for the nth element of the sequence.)

Problems

(1) Use power series inversion to show that

$$Z^{-1}\left\{\frac{1}{(z-a^T)^2}\right\} = \frac{(t-T)}{T}\,a^{(t-2T)}\,u(t-T),\ t = 0, T, 2T, 3T, \ldots\ldots$$

Check the result of Example 6.8 by using partial fractions and transform tables to find a general formula for $f(nT)$.

(2) Without using partial fractions find the inverse z-transform $\{f_n\}$ of the following $F(z)$.

(i) $\dfrac{z^2+1}{z^2-1}$,   (ii) $\dfrac{2z}{(z+1)(z+3)}$,

(iii) $\dfrac{z^2+1}{(z-1)^3}$,   (iv) $\dfrac{z}{z^2+1}$.

(3) Given that the following $F(z)$ were obtained by transforming $f(t)$, $t = 0, T,$ $2T, 3T, \ldots$, find the values $f(0), f(T), f(2T), f(3T)$ in each case.

(i) $\dfrac{z^3+3z^2+4z+1}{(z-2)^3}$,   (ii) $\dfrac{z^2+1}{(z+1)(z+2)^2}$,

(iii) $e^{1/z}$,   (iv) $\left(1+\dfrac{1}{z}\right)^7$.

(4) Use power series inversion to check the solutions of **Problem 4**, section 6.2, for $n = 0$, 1, 2, 3.

(5) Use long division to check the solution of **Problem 5** in section 6.2, for $n = 0$, 1, 2, 3.

Answers to problems

(2) (i) $f_0 = 1$; $f_n = 1 + (-1)^n$, $n = 1, 2, 3, \ldots$

(ii) $f_n = (-1)^n - (-3)^n$, $n = 0, 1, 2, 3, \ldots$

(iii) $f_0 = 0$; $f_n = n^2 - n + 1$, $n = 1, 2, 3, \ldots$

(iv) $f_n = \sin\dfrac{n\pi}{2}$, $n = 0, 1, 2, 3, \ldots$

(3) (i)   1, 9, 46, 177.

(ii) 0, 1, -5, 18.

(iii) $1, 1, \dfrac{1}{2}, \dfrac{1}{6}$.

(iv) 1, 7, 21, 35.

CHAPTER 7

Analysis of simple sampled-data systems

7.1 INTRODUCTION

In Chapter 5 we defined a sampled-data system as one where a sampled continuous function was input to a continuous processor. The present chapter applies z-transform techniques to such systems where the sampler is either an ideal sampler or a sample/hold device. The analysis is extended to deal with digital systems where the output sequence is reconstructed by a zero-order hold device and used as the input to a continuous system.

7.2 TWO IMPORTANT THEOREMS

Theorem 7.1
For the system with an ideal sampler shown in Fig. 7.1, we have

$$Z\{y(t)\} = G(z) \cdot Z\{x(t)\}$$

where $G(z) = Z\{g(t)\}, g(t) = L^{-1}\{G(s)\}$.

This result was stated in Chapter 5; the proof is now given.

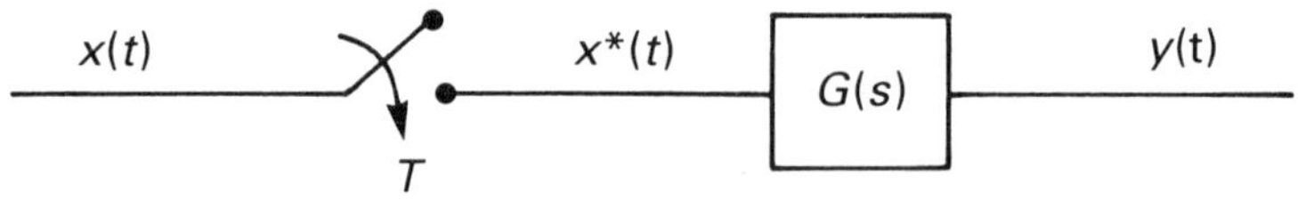

Fig. 7.1

Proof
$g(t) = L^{-1}\{G(s)\}$ is the impulse response function for the continuous system with transfer function $G(s)$, that is,

$$\delta(t) \longrightarrow \boxed{G(s)} \longrightarrow g(t)$$

and

$$\delta(t-nT) \longrightarrow \boxed{G(s)} \longrightarrow g(t-nT)u(t-nT).$$

Since $x^*(t) = \displaystyle\sum_{n=0}^{\infty} x(nT)\delta(t-nT)$

then $\quad y(t) = \sum_{n=0}^{\infty} x(nT)g(t-nT)u(t-nT)$, that is,

$$y(t) = \sum_{l=0}^{\infty} x(lT)g(t-lT),$$

simply replacing n by l and dropping $u(t-nT)$ since $g(t-nT) = 0$ for $t - nT < 0$.

Therefore $Z\{y(t)\} = \sum_{n=0}^{\infty} \left(\sum_{l=0}^{\infty} x(lT)g(nT-lT) \right) z^{-n}$

$$= \sum_{l=0}^{\infty} x(lT) \left(\sum_{n=0}^{\infty} g([n-l]T)z^{-n} \right),$$

changing the order of summation. Now put $k = n - l$ and

$$Z\{y(t)\} = \sum_{l=0}^{\infty} x(lT) \left(\sum_{k=-l}^{\infty} g(kT)z^{-k} \right) z^{-l}$$

$$= \sum_{l=0}^{\infty} x(lT) \left(\sum_{k=0}^{\infty} g(kT)z^{-k} \right) z^{-l}$$

since $g(kT) = 0$ for $kT < 0$.

Writing the above double summation as the product of two simple summations we obtain

$$Z\{y(t)\} = \sum_{k=0}^{\infty} g(kT)z^{-k} \cdot \sum_{l=0}^{\infty} x(lt)z^{-l}$$

$$= Z\{g(t)\} \cdot Z\{x(t)\}$$
$$= G(z) \cdot Z\{x(t)\}$$

which is the required result.

The *sequence* $\{y(nT)\}$ can now be found by taking the inverse z-transform. It should be emphasized that the *function* $y(t)$ cannot be obtained using z-transforms.

(Note that $y(nT) = \sum_{l=0}^{n} x(lT)g([n-l]T)$.

Therefore writing $Z\{x(t)\}$ as $X(z)$ we have

$$Z^{-1}\{G(z)X(z)\} = \left\{ \sum_{l=0}^{n} x(lt)g([n-l]T) \right\},$$

which is the z-transform **convolution theorem**).

Theorem 7.2
The system with a sample/hold shown in Fig. 7.2(a) is equivalent to the ideal

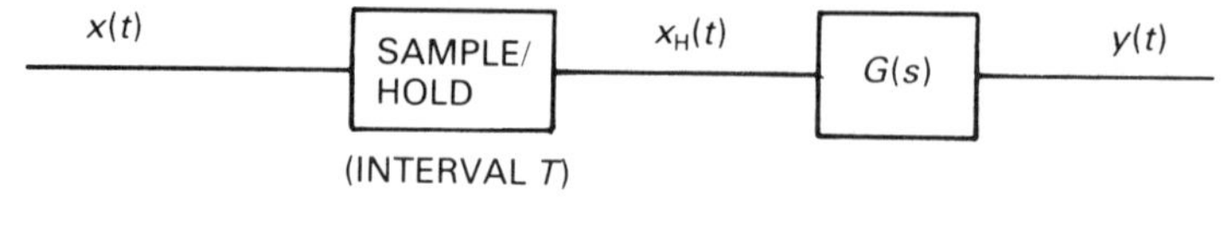

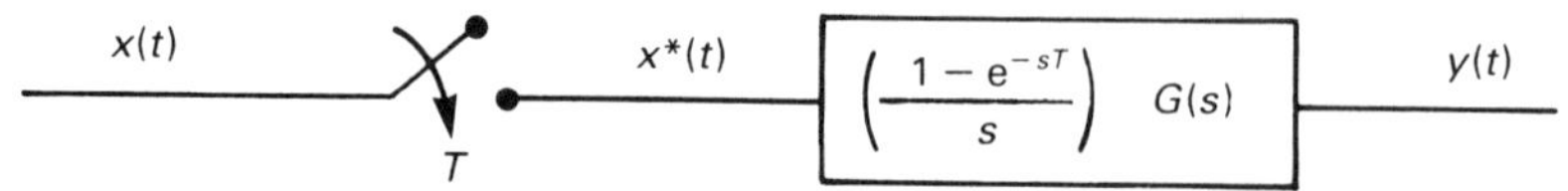

Fig. 7.2 Equivalent systems.

sampler system shown in Fig. 7.2(b). Analysis can then be carried out, as in Theorem 7.1, to obtain the output sequence $y(nT)$.

The proof of Theorem 7.2 is a simple extension of that given in example 5.2.

The output sequence $\{y(nT)\}$ obtained by sampling the continuous output $y(t)$ may be obtained as follows:

Let
$$f(t) = L^{-1}\left\{\left(\frac{1-e^{-sT}}{s}\right)G(s)\right\} \tag{7.1}$$

and
$$F(z) = Z\{f(t)\}. \tag{7.2}$$

Then Theorem 7.1 is applied to give

$$Z\{y(nT)\} = F(z)\cdot Z\{x(t)\}. \tag{7.3}$$

A more convenient form of equation (7.3) is obtained by writing

$$f_1(t) = L^{-1}\left\{\frac{G(s)}{s}\right\} \tag{7.4}$$

and
$$F_1(z) = Z\{f_1(t)\}. \tag{7.5}$$

Equations (7.1) and (7.2) now become

$$f(t) = L^{-1}\left\{\frac{G(s)}{s}\right\} - L^{-1}\left\{e^{-sT}\,\frac{G(s)}{s}\right\}$$

$$= f_1(t) - f_1(t-T)u(t-T),$$

and
$$F(z) = F_1(z) - z^{-1}F_1(z)$$

$$= \left(\frac{z-1}{z}\right)F_1(z).$$

Therefore $\{y(nT)\}$ is obtained from

$$Z\{y(nT)\} = \left(\frac{z-1}{z}\right) F_1(z) \cdot Z\{x(t)\}. \tag{7.6}$$

The application of equations (7.4), (7.5), and (7.6) is demonstrated in Example 7.3.

Example 7.1
Find the general formula for $y(nT)$ for the system shown in Fig. 7.3. (Compare with Example 5.1).

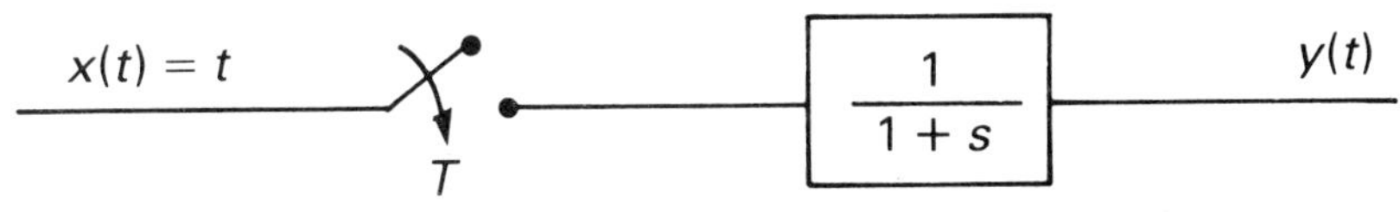

Fig. 7.3

$G(s) = \dfrac{1}{1+s}$, therefore $g(t) = e^{-t}$

and $G(z) = \dfrac{z}{z - e^{-T}}.$

$$Z\{x(t)\} = \frac{Tz}{(z-1)^2}.$$

Therefore $Z\{y(t)\} = \dfrac{z^2 T}{(z-1)^2 (z - e^{-T})}$ (from Theorem 7.1)

$$= Tz \left(\frac{(1 - e^{-T})^{-1}}{(z-1)^2} - \frac{e^{-T}(e^{-T}-1)^{-2}}{(z-1)} + \frac{e^{-T}(e^{-T}-1)^{-2}}{(z - e^{-T})} \right)$$

$$= \left(\frac{1}{1 - e^{-T}} \right) \frac{zT}{(z-1)^2} - \frac{Te^{-T}}{(1 - e^{-T})^2} \left(\frac{z}{(z-1)} - \frac{z}{(z - e^{-T})} \right).$$

Therefore $y(t) = \dfrac{t}{1 - e^{-T}} - \dfrac{Te^{-T}(1 - e^{-t})}{(1 - e^{-T})^2}$ for $t = 0, T, 2T, \ldots,$

that is $\quad y(nT) = \dfrac{nT}{1 - e^{-T}} - \dfrac{Te^{-T}(1 - e^{-nT})}{(1 - e^{-T})^2}$ for $n = 0, 1, 2, 3, \ldots$

(An alternative form is obtained after some manipulation,

$$y(nT) = T \sum_{i=0}^{n-1} (n - i) e^{-iT}.$$

Although this form is apparently simpler, it is not convenient for calculation of $y(nT)$ when n is large.)

Example 7.2

For the system in Example 7.1, obtain the function $y(t)$ for *all* t. Confirm the result for $y(nT)$ in Example 7.1.

The input to the transfer function $\dfrac{1}{1+s}$ is $x^*(t)$ where $x(t) = t$.

That is, $x^*(t) = \displaystyle\sum_{n=0}^{\infty} nT\delta(t - nT)$.

Taking Laplace transforms

$$L\{x^*(t)\} = \sum_{n=0}^{\infty} nT\,e^{-snT}.$$

Therefore $L\{y(t)\} = \displaystyle\sum_{n=0}^{\infty} nT\dfrac{e^{-snT}}{1+s},$

that is, $\qquad y(t) = \displaystyle\sum_{n=0}^{\infty} nT e^{-(t-nT)}\, u(t - nT).$

(This result can be obtained immediately using the impulse response of $\dfrac{1}{1+s}$ which is e^{-t}, that is, the response to $\delta(t - nT)$ is $e^{-(t-nT)}\, u(t - nT)$.)

By using a different letter l for the summation and then putting $t = nT$ we obtain

$$y(nT) = \sum_{l=0}^{\infty} lT e^{-(n-l)T} u(n - l)$$

$$= \sum_{l=0}^{n} lT e^{-(n-l)T}.$$

Now put $i = n - l$ and $y(nT) = T\displaystyle\sum_{i=0}^{n-1}(n-i)e^{-iT},$

which confirms the result of Example 7.1.

Example 7.3

For the system shown in Fig. 7.4 find

(i) $y(t)$ for $t = 0, T, 2T, 3T, \ldots$, using z-transforms.
(ii) $y(t)$ for all $t > 0$.

Compare the results for $t = 0, T, 2T$.

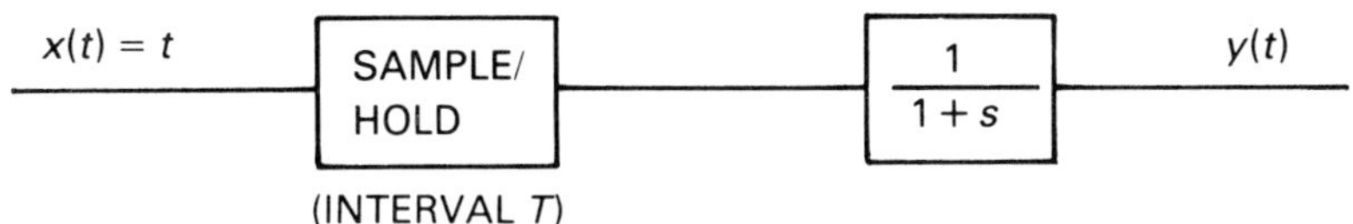

Fig. 7.4

(i) From equations (7.4) and (7.5)

$$f_1(t) = L^{-1}\left\{\frac{1}{s(s+1)}\right\} = 1 - e^{-t},$$

$$F_1(z) = \frac{z}{z-1} - \frac{z}{z-e^{-T}}.$$

Now $Z\{x(t)\} = \dfrac{zT}{(z-1)^2}$ and equation (7.6) gives

$$Z\{y(nT)\} = \left(\frac{z-1}{z}\right)\left(\frac{z}{z-1} - \frac{z}{z-e^{-T}}\right)\frac{zT}{(z-1)^2}$$

$$= \frac{zT}{(z-1)^2} - \frac{zT}{(z-1)(z-e^{-T})}$$

$$= \frac{zT}{(z-1)^2} - \left(\frac{T}{1-e^{-T}}\right)\left(\frac{z}{z-1} - \frac{z}{z-e^{-T}}\right).$$

Therefore $y(t) = t - \dfrac{T(1-e^{-t})}{1-e^{-T}}$ for $t = 0, T, 2T, 3T, \ldots$

(ii) The output of the sample/hold, $x_H(t)$, is a 'staircase' function given by

$$x_H(t) = T(u(t-T) + u(t-2T) + u(3-T) + \ldots)$$

and $L\{x_H(t)\} = \dfrac{T}{s}(e^{-sT} + e^{-s2T} + e^{-s3T} + \ldots).$

Therefore $L\{y(t)\} = \dfrac{T}{s(s+1)}(e^{-sT} + e^{-s2T} + e^{-s3T} + \ldots),$

that is, $y(t) = T\big((1 - e^{-(t-T)})u(t-T) + (1 - e^{-(t-2T)})u(t-2T)$

$$+ (1 - e^{-(t-3T)})u(t-3T) + \ldots\big)   \text{for } t > 0.$$

Comparing the results:

for $t = 0$ (i) gives $y = 0$
                   (ii) gives $y = 0,$

for $t = T$ (i) gives $y = T - T = 0$
                   (ii) gives $y = T(1 - 1) = 0,$

for $t = 2T$ (i) gives $y = 2T - T\left(\dfrac{1 - e^{-2T}}{1 - e^{-T}} \right)$

$$= 2T - T(1 + e^{-T})$$
$$= T - Te^{-T},$$

(ii) gives $y = T(1 - e^{-T} + 1 - 1)$
$$= T - Te^{-T}.$$

Problems

(1) For the system shown in Fig. 7.5 show that

$$y(nT) = \frac{1 - e^{-(n+1)T}}{1 - e^{-T}}, \quad n = 0, 1, 2, 3, \ldots$$

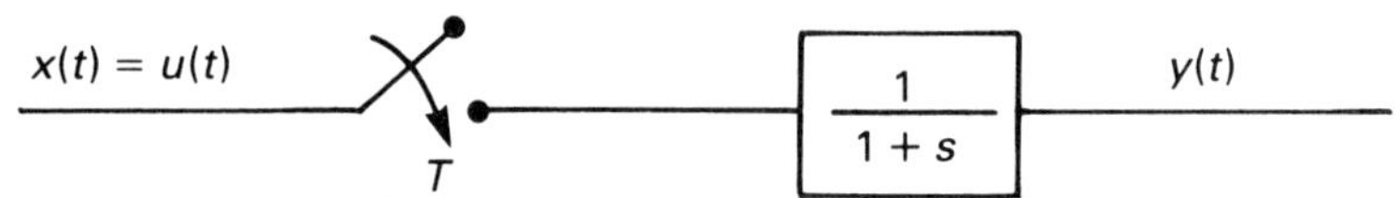

Fig. 7.5

(2) If in Problem 1 the simple lag $\dfrac{1}{1 + s}$ is replaced by an integrator $\dfrac{1}{s}$, use
z-transforms to show that $y(nT) = n + 1$. What is $y(t)$ for all t?

(3) Given the system of Fig. 7.6, use z-transforms to calculate $y(0)$, $y(T)$, $y(2T)$, $y(3T)$, $y(4T)$, $y(5T)$, and $y(6T)$ for the cases

(i) $T = \dfrac{\pi}{2}$

(ii) $T = \pi.$

Check your answers by finding $y(t)$ for all t.

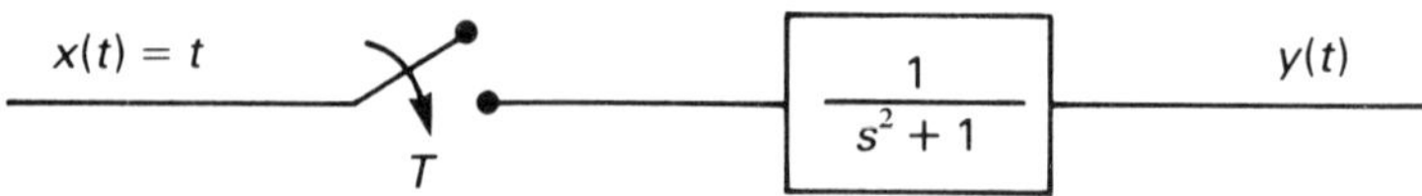

Fig. 7.6

(4) Confirm that for $T = \dfrac{\pi}{4}$

$$Z\{e^{-t}\cos 2t\} = \frac{z^2}{z^2 + e^{-\pi/2}},$$

$$Z\{e^{-t}\sin 2t\} = \frac{ze^{-\pi/4}}{z^2 + e^{-\pi/2}}.$$

For the system shown in Fig. 7.7, show that
(i) when $x(t) = u(t)$

$$y\left(\pi\frac{\pi}{4}\right) = \frac{e^{-\pi/2}}{1 + e^{-\pi/2}}\left(e^{\pi/2} + e^{-n\pi/4}\cos\frac{n\pi}{2} + e^{(1-n)\pi/4}\sin\frac{n\pi}{2}\right),$$

$$n = 0, 1, 2, 3, \ldots,$$

(ii) when $x(t) = t$

$$y(0) = 0, \ y\left(\frac{\pi}{4}\right) = \frac{\pi}{4}, \ y\left(\frac{\pi}{2}\right) = \frac{\pi}{2},$$

$$y\left(\frac{3\pi}{4}\right) = \frac{\pi}{4}(3 - e^{-\pi/2}), \ y(\pi) = \frac{\pi}{4}(4 - 2e^{-\pi/2}),$$

$$y\left(\frac{5\pi}{4}\right) = \frac{\pi}{4}(5 - 3e^{-\pi/2} + e^{-\pi}),$$

$$y\left(\frac{3\pi}{2}\right) = \frac{\pi}{4}(6 - 4e^{-\pi/2} + 2e^{-\pi}).$$

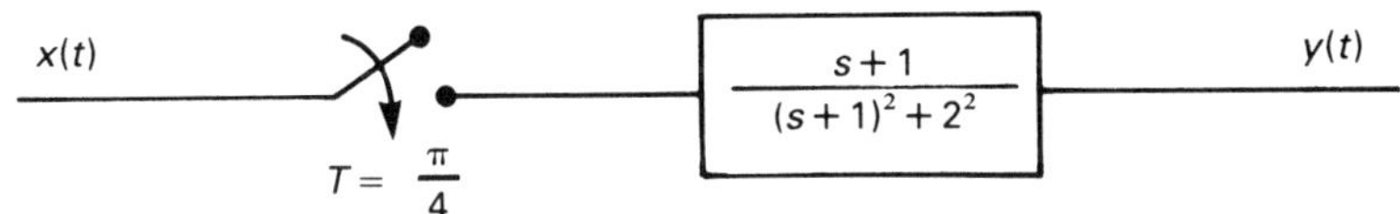

Fig. 7.7

(5) Given the system of Fig. 7.8, use z-transforms to evaluate $y(t)$ for $t = 0, 1, 2, 3, 4, 5$. Check your results by finding $y(t)$ for all t.

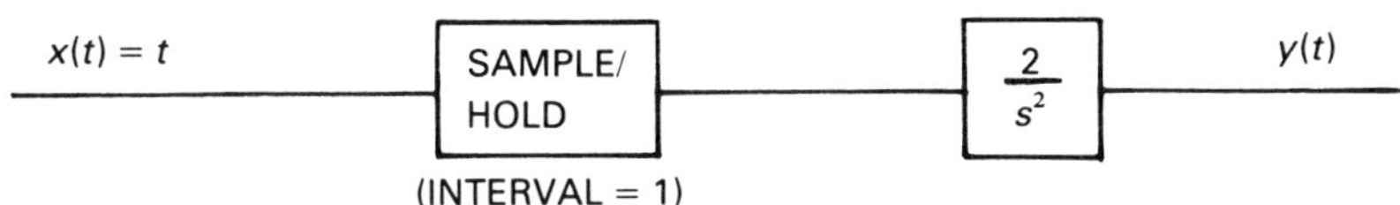

Fig. 7.8

(6) Find $y(nT)$, $n = 0, 1, 2, 3, \ldots$, for the system shown in Fig. 7.9.

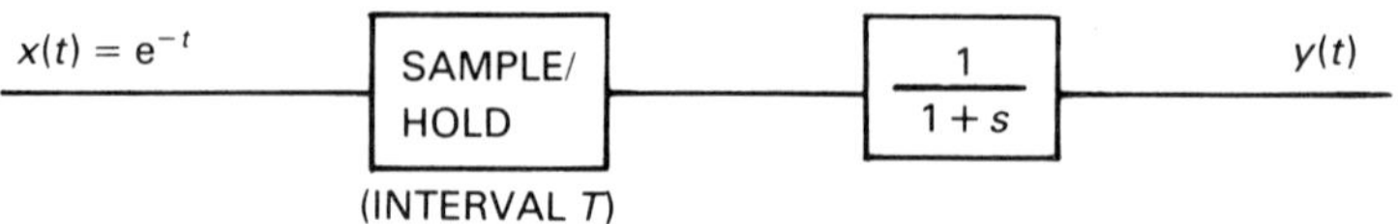

Fig. 7.9

(7) For the system shown in Fig. 7.10 show that

$$y\left(\frac{n\pi}{2}\right) = \beta^2 \alpha \left(e^{-n\pi} - \cos\frac{n\pi}{2}\right) + \alpha \sin\frac{n\pi}{2}.$$

$n = 0, 1, 2, 3, \ldots,$

where $\beta = e^{-\pi/2}$, $\alpha = \dfrac{2}{\beta^4 + 1}$.

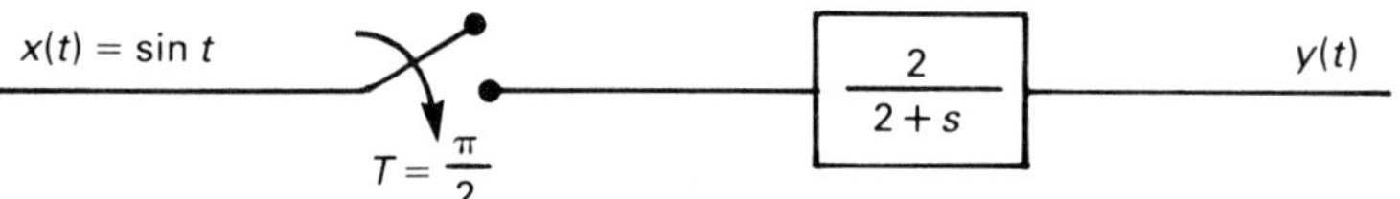

Fig. 7.10

Answers to problems

(2) $y(t) = \displaystyle\sum_{l=0}^{\infty} u(t - lT)$

(3) (i) $0, 0, \dfrac{\pi}{2}, \pi, \pi, \pi, \dfrac{3\pi}{2};$

$$y(t) = \sum_{l=0}^{\infty} \frac{l\pi}{2} \sin\left(t - \frac{l\pi}{2}\right) u\left(t - \frac{l\pi}{2}\right).$$

(ii) $0, 0, 0, 0, 0, 0, 0;$

$$y(t) = \sum_{l=0}^{\infty} l\pi \sin(t - l\pi)\, u(t - l\pi).$$

(5) $0, 0, 1, 5, 14, 30;$

$$y(t) = \sum_{l=1}^{\infty} (t - l)^2 u(t - l).$$

(6) $y(nT) = n(e^{-(n-1)T} - e^{-nT}).$

7.3 EXTENSION TO DIGITAL SYSTEMS

It is frequently necessary to **reconstruct** the output sequence of a digital processor to form a continuous function which can then be the input to a continuous processor. The simplest reconstruction device is the **zero-order hold** (a type of DAC, digital-to-analog converter) which outputs a 'staircase' function, as shown in Fig. 7.11.

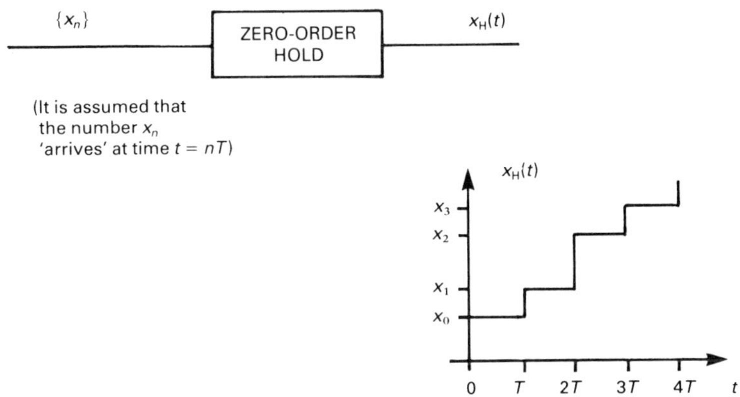

Fig. 7.11 Zero-order hold reconstruction device.

If a function $x(t)$, such that $x(nT) = x_n$, is input to a sample/hold device, then the output is the function $x_H(t)$ as shown in Fig. 7.11. It follows that the two systems shown in Figs 7.12(a) and 7.12(b) produce the *same output* $y(t)$ when $x(nT) = x_n$. Therefore in *both* cases equation (7.6) applies:

that is, $Z\{y(t)\} = \left(\dfrac{z-1}{z}\right) F_1(z) \cdot Z\{x(t)\}$,

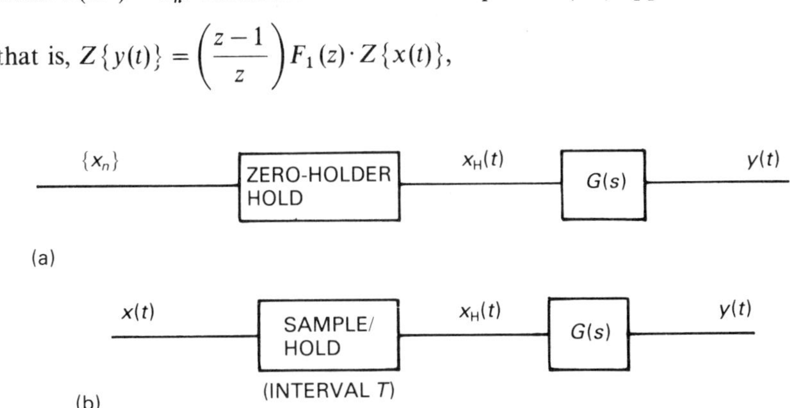

Fig. 7.12 System (a) and system (b) have the same output when $\{x_n\}$ is a time series such that $x_n = x(nT)$.

where $\quad F_1(z) = Z\left\{L^{-1}\left\{\dfrac{G(s)}{s}\right\}\right\}.$

The *sequence* $\{y(nT)\}$ is obtained by taking the inverse z-transform. (Note $Z\{x(t)\} = Z\{x_n\}$ since $x(nT) = x_n$.)

Example 7.4

The sequence $\{x_n\}$ is output from a digital processor such that x_n appears at time $t = nT$. Find $y(nT)$ for the system given in Fig. 7.13 when $x_n = n$.

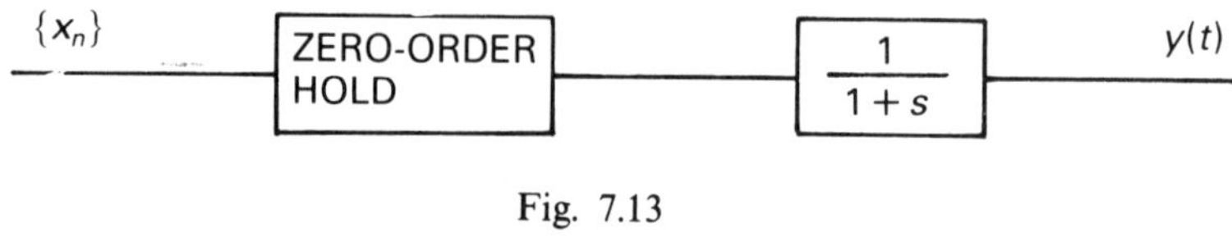

Fig. 7.13

Following the procedure of equations (7.4), (7.5) and (7.6),

$$f_1(t) = L^{-1}\left\{\frac{1}{s(s+1)}\right\} = 1 - e^{-t},$$

$$F_1(z) = \frac{z}{z-1} - \frac{z}{z-e^{-T}},$$

$$Z\{x_n\} = Z\{n\} = \frac{z}{(z-1)^2}.$$

Therefore $\quad Z\{y(t)\} = \left(\dfrac{z-1}{z}\right)\left(\dfrac{z}{z-1} - \dfrac{z}{z-e^{-T}}\right)\dfrac{z}{(z-1)^2}$

$$= \frac{z}{(z-1)^2} - \frac{z}{(z-1)(z-e^{-T})}.$$

Therefore $y(t) = \dfrac{t}{T} - \dfrac{1-e^{-t}}{1-e^{-T}},\quad$ (see Example 7.3), for $t = 0, T, 2T, 3T, \ldots,$

that is, $y(nT) = n - \dfrac{1-e^{-nT}}{1-e^{-T}}$ for $n = 0, 1, 2, 3, \ldots.$

Example 7.5

Consider the system shown in Fig. 7.14. The digital processor is programmed to execute the difference equation, $c_{n+1} = f_{n+1} + f_n + 0.5c_n$; $c_0 = 0$, and the sampling interval is T for both samplers.

Given that $x(t) = e^{-t}$ find y_0, y_1, y_2 and y_3.

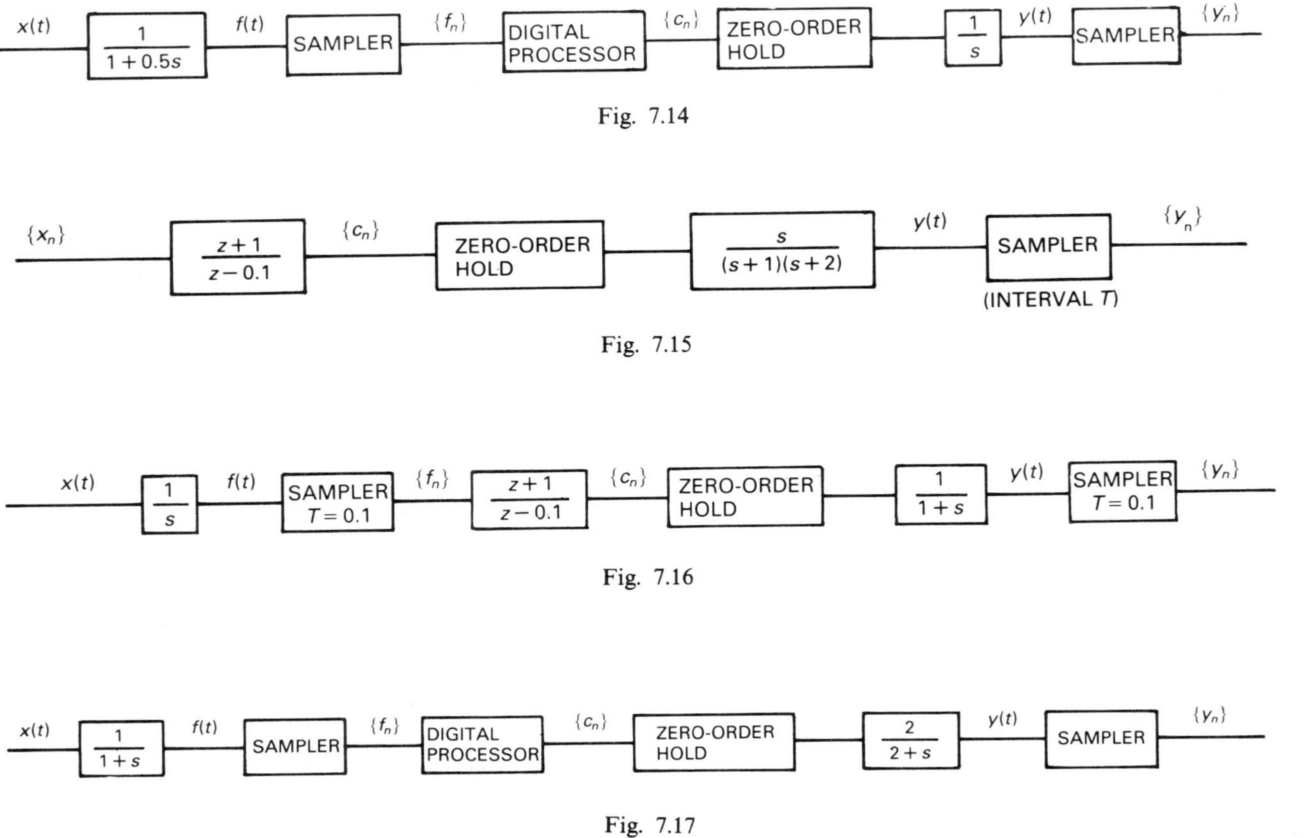

Fig. 7.14

Fig. 7.15

Fig. 7.16

Fig. 7.17

$$L\{f(t)\} = \frac{1}{(1+s)(1+0.5s)} = \frac{2}{1+s} - \frac{1}{1+0.5s}.$$

Therefore $f(t) = 2e^{-t} - 2e^{-2t}$

and $\quad Z\{f_n\} = \dfrac{2z}{z - e^{-T}} - \dfrac{2z}{z - e^{-2T}}.$

Transforming the difference equation gives

$$zZ\{c_n\} = (z+1)Z\{f_n\} + 0.5\,Z\{c_n\}$$

since $c_0 = f_0 = 0$.

Therefore $\dfrac{Z\{c_n\}}{Z(f_n)} = \dfrac{z+1}{z-0.5}$

and $\quad Z\{c_n\} = \dfrac{2z(z+1)}{(z - e^{-T})(z - 0.5)} - \dfrac{2z(z+1)}{(z - e^{-2T})(z - 0.5)}.$

Take $f_1(t) = L^{-1}\left\{\dfrac{1}{s^2}\right\} = t$

and hence $F_1(z) = \dfrac{zT}{(z-1)^2}.$

Therefore $Z\{y_n\} = Z\{y(t)\}$ is given by

$$Z\{y_n\} = \left(\frac{z-1}{z}\right)\frac{zT}{(z-1)^2}Z\{c_n\}$$

$$= \frac{2z(z+1)T}{(z - e^{-T})(z - 0.5)(z - 1)} - \frac{2z(z+1)T}{(z - e^{-2T})(z - 0.5)(z - 1)}$$

$$= \frac{2T(z^{-1}+z^{-2})}{(1 - 0.5z^{-1})(1 - z^{-1})}\left(\frac{1}{1 - e^{-T}z^{-1}} - \frac{1}{1 - e^{-2T}z^{-1}}\right)$$

$$= 2T(z^{-1}+z^{-2})(1 + 0.5z^{-1} + 0.25z^{-2} + \ldots)(1 + z^{-1} + z^{-2} + \ldots)$$

$$\cdot\left((e^{-T} - e^{-2T})z^{-1} + (e^{-2T} - e^{-4T})z^{-2} + \ldots\right).$$

Evaluating coefficients of powers of z, we get

$$y_0 = 0,\; y_1 = 0,\; y_2 = 2T(e^{-T} - e^{-2T}),$$
$$y_3 = 2T(e^{-T} - e^{-2T})(1 + 0.5 + 1) + 2T(e^{-2T} - e^{-4T})$$
$$= T(5e^{-T} - 3e^{-2T} - 2e^{-4T}).$$

(A general result for y_n can be obtained by expressing $Z\{y_n\}$ in partial fraction form and using z-transform tables.)

Problems

(1) For the system of Fig. 7.15 find the sequence $\{y_n\}$, given that $\{x_n\}$ is the unit step sequence $\{u_n\}$. The numbers $x_0, x_1, x_2, x_3, \ldots$ appear at times $t = 0, T, 2T, 3T, \ldots$.

(2) Find y_0, y_1, y_2, and y_3 for the system shown in Fig. 7.16 when $x(t)$ is the unit step function $u(t)$.

(3) The digital processor in Fig. 7.17 executes the difference equation:

$$c_{n+1} = f_n + 0.1\,c_n, \quad c_0 = 0.$$

Find y_0, y_1, y_2, and y_3 when $x(t) = u(t)$. The sampling interval is T for both samplers.

Answers to problems

(1) $y_n = \left(\dfrac{1}{e^{-T} - 0.1}\right)\left((e^{-T} + 1)e^{-nT} - 1.1(0.1)^n\right)$

$\qquad - \left(\dfrac{1}{e^{-2T} - 0.1}\right)\left((e^{-2T} + 1)e^{-2nT} - 1.1(0.1)^n\right)$

$\qquad$ or $\; y_n = e^{-nT} - e^{-2nT} + 1.1 \displaystyle\sum_{i=1}^{n} (0.1)^{i-1}(e^{-(n-i)T} - e^{-(n-i)2T})$:

(2) $y_0 = y_1 = 0, \; y_2 = 0.1(1 - e^{-0.1}),$
$\qquad y_3 = 0.31 - 0.21e^{-0.1} - 0.1e^{-0.2}.$

(3) $y_0 = y_1 = y_2 = 0, \; y_3 = 1 - e^{-T} - e^{-2T} + e^{-3T}.$

CHAPTER 8

Digital filters

8.1 INTRODUCTION

The concept of filtering, that is, shaping the frequency spectrum of an input signal, has its origin in the theory of linear continuous systems. A great deal of information is available for the design of a continuous filter (or **analog filter**) to meet a particular requirement. In the present chapter we consider the problem of replacing a given analog filter by a suitably programmed digital computer operating on the sampled input. The motivation for this is that the cost of digital implementation is often considerably lower than that of its analog equivalent. A small special purpose computer can be time-shared among several components of a complex system, and accuracy is dependent only on the computer word length. Also the filtering operation can be modified if necessary simply by changing the computer program, in contrast to the analog filter which may require to be physically rebuilt.

8.2 DIGITAL FILTER DESIGN

Fig. 8.1(a) shows the analog filter $G(s)$ together with its input and output, and Fig. 8.1(b) shows its digital equivalent.

The transfer function $D(z)$ describes the operation of the digital computer, and is referred to as the **digital filter**. The object of this chapter is to design the digital filter $D(z)$ so that the reconstructed output $y_H(t)$ is as close as possible to

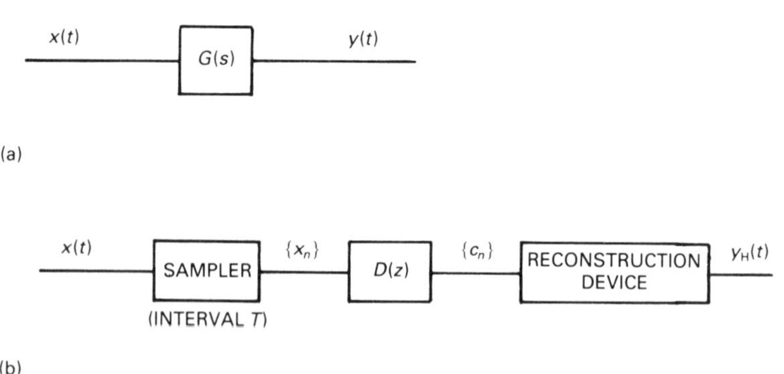

Fig. 8.1 (a) analog filter, (b) equivalent digital filter.

the analog filter output, $y(t)$. Once $D(z)$ is known, the digital computer can be programmed to execute the corresponding linear difference equation.

8.3 THE INPUT-INVARIANT TECHNIQUE

It is a simple matter to design $D(z)$ such that the digital computer output sequence, $\{c_n\}$, satisfies $c_n = y(nT)$ for all n *provided that the input function, $x(t)$, is known.*

The procedure is as follows:

(1) Calculate $y(t)$ from $y(t) = L^{-1}\{X(s)G(s)\}$.
(2) Obtain the z-transforms $X(z) = Z\{x(t)\}$ and $Y(z) = Z\{y(t)\}$.
(3) We require that $c_n = y(nT)$, that is, $Z\{c_n\} = Z\{y(t)\}$.
     Therefore we take $D(z) = Y(z)/X(z)$ so that $Z\{c_n\} = X(z)\cdot D(z) = Y(z)$.

If the reconstruction device is a zero-order hold, then the outputs $y(t)$ and $y_H(t)$ compare as in Fig. 8.2. Note that $y_H(t)$ lags behind $y(t)$ by an average amount $T/2$. This may not be significant if T is small; otherwise, some advance needs to be built into the system.

The main disadvantage of the above technique is that $c_n = y(nT)$ exactly *only* for the particular input under consideration. The impulse-invariant design may give inaccurate output when the input is a step function (see Example 8.1). Modifications to the technique are available to deal with inputs which are linear combinations of functions of the same type, see Stearns (1975). These are not included in this book, but the next section considers techniques which attempt to deal with general inputs.

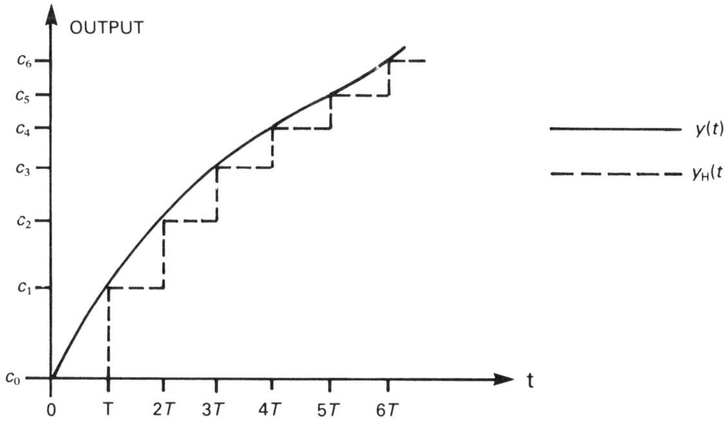

Fig. 8.2. Input-invariant technique with zero-order hold reconstruction.

Example 8.1

Design an impulse-invariant digital filter to be equivalent to the analog filter $\dfrac{1}{1+s}$. Compare the outputs of the analog and digital systems when the input is a unit step function.

The equivalent systems are shown in Fig. 8.3.

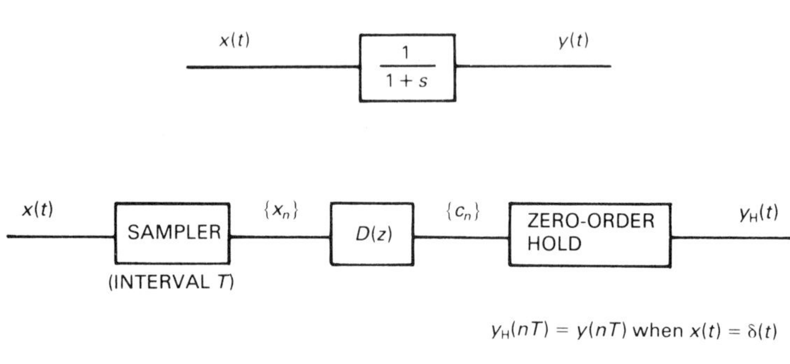

Fig. 8.3

For $x(t) = \delta(t)$, the unit impulse function, $L\{x(t)\} = 1$, and $L\{y(t)\} = \dfrac{1}{1+s}$.

Therefore $y(t) = e^{-t}$.

To analyse the digital system, approximate to the unit impulse input by taking

$$x(t) = \begin{cases} \dfrac{1}{T} & \text{for } 0 \leqslant t < T \\ 0 & \text{for } t \geqslant T \end{cases}$$

Therefore $x_0 = \dfrac{1}{T}$, $x_n = 0$ for $n = 1, 2, 3, \ldots$,

and $Z\{x_n\} = \dfrac{1}{T}$.

Also $Z\{y(t)\} = \dfrac{z}{z - e^{-T}}$.

We require $D(z) = \dfrac{Z\{y(t)\}}{Z\{x(t)\}} = \dfrac{Tz}{z - e^{-T}}$.

(We can check that $c_n = y(nT)$ for all n, since

$$Z\{c_n\} = D(z) \cdot Z\{x_n\} = \frac{z}{z - e^{-T}}.$$

Therefore $c_n = e^{-nT} = y(nT)$.)

Now compare the outputs of the continuous system and the impulse-invariant digital system when the input is a unit step function.

That is, $x(t) = u(t)$, $L\{x(t)\} = \dfrac{1}{s}$, $Z\{x(t)\} = \dfrac{z}{z-1}$.

For the analog system,

$$L\{y(t)\} = \frac{1}{s(1+s)} = \frac{1}{s} - \frac{1}{1+s}$$

and $y(t) = 1 - e^{-t}$.

For the impulse-invariant system,

$$Z\{c_n\} = \left(\frac{z}{z-1}\right)\left(\frac{Tz}{z-e^{-T}}\right)$$

$$= \frac{T}{(1-e^{-T})}\left(\frac{z}{z-1} - \frac{e^{-T}z}{z-e^{-T}}\right).$$

Therefore $c_n = \dfrac{T}{(1-e^{-T})}\left(1 - e^{-T}e^{-nT}\right)$

$$= \frac{T(1-e^{-(n+1)T})}{1-e^{-T}}.$$

Comparing with $y(nT) = 1 - e^{-nT}$,

$$y(0) = 0 \qquad\qquad c_0 = T$$
$$y(T) = 1 - e^{-T} \qquad\qquad c_1 = T(1+e^{-T})$$
$$Y(2T) = 1 - e^{-2T} \qquad\qquad c_2 = T(1+e^{-T}+e^{-2T})$$

$$\lim_{n \to \infty} y(nT) = 1 \qquad\qquad \lim_{n \to \infty} c_n = \frac{T}{1-e^{-T}}.$$

For example if $T = 0.2$,

$$y(0) = 0 \qquad\qquad c_0 = 0.2$$
$$y(0.2) = 0.1813 \qquad\qquad c_1 = 0.3637$$
$$y(0.4) = 0.3297 \qquad\qquad c_2 = 0.4978$$

$$\lim_{n \to \infty} y(0.2n) = 1 \qquad\qquad \lim_{n \to \infty} c_n = 1.1033.$$

This example illustrates that the impulse-invariant digital filter for $\dfrac{1}{1+s}$ does not perform particularly well for a step input, unless T is small.

   (Problem 1 suggests a method for modifying the impulse-invariant digital filter so that it gives more accurate results for a step input.)

Example 8.2

Design a step-invariant digital filter to be equivalent to the analog filter $\dfrac{1}{1+s}$. Write down the difference equation to be executed by the digital computer.

$$x(t) = u(t), \quad L\{x(t)\} = \frac{1}{s}$$

Therefore $L\{y(t)\} = \dfrac{1}{s(s+1)} = \dfrac{1}{s} - \dfrac{1}{1+s}$

and $\qquad y(t) = 1 - e^{-t}.$

$$Z\{x(t)\} = \frac{z}{z-1}$$

$$Z\{y(t)\} = \frac{z}{z-1} - \frac{z}{z-e^{-T}}.$$

Therefore take $D(z) = \dfrac{Z\{y(t)\}}{Z\{x(t)\}} = 1 - \dfrac{z-1}{z-e^{-T}}$

$$= \frac{1-e^{-T}}{z-e^{-T}}.$$

To obtain the corresponding difference equation write

$$\frac{Z\{c_n\}}{Z\{x_n\}} = D(z) = \frac{z^{-1}(1-e^{-T})}{1-z^{-1}e^{-T}}.$$

Therefore $c_n - e^{-T}c_{n-1} = (1-e^{-T})x_{n-1}$,

that is, $c_n = (1-e^{-T})x_{n-1} + e^{-T}c_{n-1}.$

(Check $c_{-1} = 0$; $x_{-1} = 0$, $x_n = 1$, $n \geqslant 0$.

$$\begin{aligned}
c_0 &= 0, & &\text{compare } y(0) = 0. \\
c_1 &= 1 - e^{-T}, & &\text{compare } y(T) = 1 - e^{-T}, \\
c_2 &= 1 - e^{-T} + e^{-T}(1 - e^{-T}) = 1 - e^{-2T}, \\
& & &\text{compare } y(2T) = 1 - e^{-2T}, \\
c_3 &= 1 - e^{-T} + e^{-T}(1 - e^{-2T}) = 1 - e^{-3T}, \\
& & &\text{compare } y(3T) = 1 - e^{-3T}, \text{ etc.)}
\end{aligned}$$

Problems

(1) Example 8.1 designed the impulse-invariant digital filter $D(z) = \dfrac{Tz}{z - e^{-T}}$ corresponding to the analog filter $\dfrac{1}{1+s}$. Modify this digital filter by using a constant scale factor so that a *step* function input produces the same steady-state output for both analog and digital systems,

(that is, $\lim\limits_{n \to \infty} c_n = \lim\limits_{n \to \infty} y(nT)$).

Compare the outputs when $T = 0.2$.

(2) Design a ramp-invariant digital filter (that is, $x(t) = t$) to be equivalent to the analog filter $\dfrac{1}{1+s}$. Compare the analog and digital outputs when the input $x(t)$ is a unit step function and $T = 0.2$.

(3) A digital filter is required to be equivalent to the analog filter $\dfrac{1}{1+s}$. Given that the input to both systems is $x(t) = \sin \omega t$ and that the outputs must be identical at the sampling times $t = 0, T, 2T, 3T, \ldots$, obtain the linear difference equation relating digital computer input and output sequences.

Answers to problems

(1) $D(z) = \dfrac{(1 - e^{-T})z}{z - e^{-T}}$

$c_0 = 0.1813,$	compare $y(0) = 0.$
$c_1 = 0.3297,$	compare $y(0.2) = 0.1813.$
$c_2 = 0.4512,$	compare $y(0.4) = 0.3297.$
$c_n = 1 - e^{-0.2(n+1)}$	compare $y(0.2n) = 1 - e^{-0.2n}.$
$\lim\limits_{n \to \infty} c_n = 1,$	compare $\lim\limits_{n \to \infty} y(0.2n) = 1.$

(2) $D(z) = \dfrac{(T - 1 + e^{-T})z - Te^{-T} + 1 - e^{-T}}{T(z - e^{-T})}.$

For $x(t) = u(t)$, $c_n = 1 - \dfrac{1}{T}(e^{-nT} - e^{-(n+1)T})$

$y(nT) = 1 - e^{-nT}.$

Comparing when $T = 0.2$,

$c_0 = 0.0937,$	$y(0) = 0.$
$c_1 = 0.2579,$	$y(0.2) = 0.1813.$
$c_2 = 0.3925,$	$y(0.4) = 0.3297.$
$\lim\limits_{n \to \infty} c_n = 1,$	$\lim\limits_{n \to \infty} y(0.2n) = 1.$

$$(3) \quad c_n = e^{-T}c_{n-1} + \frac{1}{(1+\omega^2)\sin \omega T}((\sin \omega T - \omega \cos \omega T + \omega e^{-T})x_n$$

$$+ (\omega - e^{-T}\sin \omega T - \omega e^{-T}\cos \omega T)x_{n-1}).$$

8.4 INPUT-INDEPENDENT TECHNIQUES

Many methods are available in the literature which attempt to overcome the difficulties of the previous section. Two such methods are considered in the present section. The requirement is that the digital filter will operate satisfactorily for a range of input functions.

(i) The trapezoidal integration technique

Consider the analog system shown in Fig. 8.4. The differential equation describing this system is

$$\frac{dy}{dt} + ay = ax, \text{ that is,}$$

$$\frac{dy}{dt} = a(x - y).$$

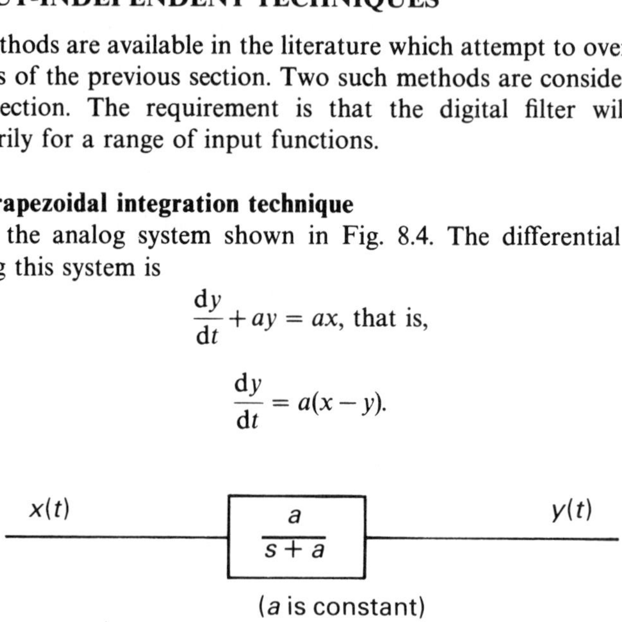

$$x(t) \qquad \boxed{\dfrac{a}{s+a}} \qquad y(t)$$

$$(a \text{ is constant})$$

Fig. 8.4

We integrate this equation between $t = (n-1)T$ and $t = nT$ using trapezoidal integration.

$$\left(\int_{(n-1)T}^{nT} f(t)dt = \frac{T}{2}(f_{n-1} + f_n) \text{ where } f_n = f(nT). \right)$$

$$\int_{(n-1)T}^{nT} dy = \int_{(n-1)T}^{nT} a(x - y)dt, \text{ that is,}$$

$$y_n - y_{n-1} = \frac{aT}{2}(x_{n-1} - y_{n-1} + x_n - y_n).$$

(This equation can be solved explicitly for y_n and is the difference equation which is an approximation to the differential equation.)

Taking z-transforms,

$$\left(1 - z^{-1} + \frac{aT}{2}(1 + z^{-1})\right) Z\{y_n\} = \frac{aT}{2}(1 + z^{-1})Z\{x_n\}.$$

Therefore $\dfrac{Z\{y_n\}}{Z\{x_n\}} = \dfrac{\dfrac{aT}{2}(z+1)}{\dfrac{aT}{2}(z+1) + (z-1)}.$

We take this transfer function to be $D(z)$, the digital filter, so that the digital output sequence $\{c_n\}$ is *approximately* $\{y_n\} = \{y(nT)\}$. Therefore, for $G(s) = \dfrac{a}{s+a}$, the trapezoidal approximation gives

$$D(z) = \frac{\dfrac{aT}{2}(z+1)}{\dfrac{aT}{2}(z+1) + (z-1)}.$$

This can be written

$$D(z) = \frac{a}{a + \dfrac{2}{T}\left(\dfrac{z-1}{z+1}\right)}.$$

Note that $D(z)$ is obtained from $G(s)$ by the substitution $s = \dfrac{2}{T}\left(\dfrac{z-1}{z+1}\right)$.

Since any general (proper) transfer function, $G(s)$, can be expressed by partial fractions as a combination of terms such as $\dfrac{a}{s+a}$, it follows that the substitution,

$$s = \frac{2}{T}\left(\frac{z-1}{z+1}\right)$$

can be applied to a *general* analog filter, $G(s)$, to obtain the approximate corresponding digital filter, $D(z)$.

Example 8.3

Use the trapezoidal approximation to design a digital filter corresponding to $G(s) = \dfrac{1}{s^2 + s + 1}$. Write down the difference equation relating output and input of the digital computer.

————————————

$$D(z) = \cfrac{1}{\left(\cfrac{2(z-1)}{T(z+1)}\right)^2 + \cfrac{2(z-1)}{T(z+1)} + 1}$$

$$= \frac{T^2(z+1)^2}{4(z-1)^2 + 2T(z^2-1) + T^2(z+1)^2}$$

$$= \frac{T^2(z^2+2z+1)}{(4+2T+T^2)z^2 + (2T^2-8)z + (4-2T+T^2)}.$$

That is, $\dfrac{Z\{c_n\}}{Z\{x_n\}} = \dfrac{T^2(1+2z^{-1}+z^{-2})}{(4+2T+T^2) + (2T^2-8)z^{-1} + (4-2T+T^2)z^{-2}}.$

Therefore the required difference equation is

$$c_n = \frac{1}{(4+2T+T^2)}\,(T^2(x_n + 2x_{n-1} + x_{n-2})$$

$$- (2T^2-8)c_{n-1} - (4-2T+T^2)c_{n-2}).$$

The trapezoidal approximation is very easy to use and it has another advantage. The stable region in the complex s-plane is mapped into the stable region in the complex z-plane by the transformation $s = \dfrac{2}{T}\left(\dfrac{z-1}{z+1}\right)$. The stable region in the $s = \sigma + j\omega$ plane is $\sigma < 0$, and in Chapter 6 it was shown that the transformation $z = e^{sT}$ maps this region into $x^2 + y^2 < 1$, which is the stable region in the $z = x + jy$ plane, see Fig. 8.5.

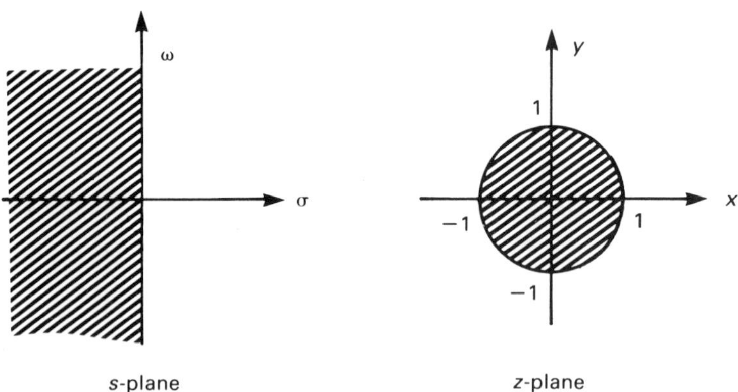

Fig. 8.5 Transformations $z = e^{sT}$ and $s = \dfrac{2}{T}\left(\dfrac{z-1}{z+1}\right)$.

It is easily proved that the trapezoidal transformation $s = \dfrac{2}{T}\left(\dfrac{z-1}{z+1}\right)$ carries out the same mapping.

We have $\sigma + j\omega = \dfrac{2}{T}\left(\dfrac{(x-1)+jy}{(x+1)+jy}\right)$, and equating real parts gives

$$\sigma = \frac{2}{T}\left(\frac{x^2+y^2-1}{(x+1)^2+y^2}\right).$$

Therefore $\sigma < 0$ corresponds to $x^2 + y^2 < 1$. It follows that, if the output of the analog system shown in Fig. 8.1(a) is stable (that is, $G(s)$ has no poles in the right-hand half-plane) then the trapezoidal approximation for $D(z)$ will produce *stable* digital computer output $\{c_n\}$, since $D(z)$ will have no poles outside the unit circle.

(ii) **The pole-zero matching technique**
This method matches poles and zeros of $G(s)$ and $D(z)$ using the relation $z = e^{sT}$. For example, if $G(s)$ has a pole at $s = -a$, then we would expect $D(z)$ to have a pole at $z = e^{-aT}$. Before detailing this method we recall the trapezoidal approximation

$$D(z) = \frac{a}{a+\dfrac{2}{T}\left(\dfrac{z-1}{z+1}\right)} \text{ to } G(s) = \frac{a}{a+s}.$$

After a little manipulation the trapezoidal $D(z)$ can be written

$$D(z) = \left(\frac{a}{a+\dfrac{2}{T}}\right)\left(\frac{z+1}{z-\left(1-\dfrac{aT}{2}\right)\Big/\left(1+\dfrac{aT}{2}\right)}\right).$$

If the sampling interval, T, is such that aT is small, we can write

$$\left(1-\frac{aT}{2}\right)\left(1+\frac{aT}{2}\right)^{-1} \simeq \left(1-\frac{aT}{2}\right)\left(1-\frac{aT}{2}\right)$$

$$\simeq 1-aT$$
$$\simeq e^{-aT}$$

Therefore $D(z) \simeq \left(\dfrac{a}{a+\dfrac{2}{T}}\right)\left(\dfrac{z+1}{z-e^{-aT}}\right)$ for the trapezoidal method.

Thus the pole at $s = -a$ in $G(s)$ shows up (as we would expect) as a pole at $z = e^{-aT}$ in $D(z)$.

We still have to explain the presence of the $z+1$ factor in the numerator.

This corresponds to the fact that the $G(s)$ has a single zero when s is infinite, and the explanation is related to the choice of sampling interval, T. We consider the effect of the filter on high-frequency inputs. The analog filter $\dfrac{a}{s+a}$ reduces towards zero input components of high frequency since

$$G(\mathrm{j}\omega) = \frac{a}{a+\mathrm{j}\omega} \to 0 \quad \text{as } \omega \to \infty.$$

It is reasonable to make the digital filter reduce to zero its maximum frequency input component. Assuming that the input $x(t)$ is band-limited with maximum frequency ω_M rad/sec, then the sampling interval T must be less than $\dfrac{\pi}{\omega_M}$ (Sampling Theorem, see Stearns (1975)). Put another way, for a given sampling interval T, the maximum frequency component of $x(t)$ must have freqency less than $\dfrac{\pi}{T}$ (Nyquist frequency). Therefore we make the digital filter reduce to zero, input with frequency $\omega = \dfrac{\pi}{T}$; that is, $D(z)$ must be zero when $s = \mathrm{j}\dfrac{\pi}{T}$ and $z = \mathrm{e}^{\mathrm{j}(\pi/T)T} = -1$.

The pole-zero matching method is carried out as follows.

(1) Any zero, $s = -b$, of $G(s)$ implies that $D(z)$ has a zero at $z = \mathrm{e}^{-bT}$.
(2) Any pole, $s = -a$, of $G(s)$ implies that $D(z)$ has a pole at $z = \mathrm{e}^{-aT}$.
(3) Any zero for s infinite implies that $D(z)$ has a zero at $z = -1$.
(4) Choose the gain of $D(z)$ to match the gain of $G(s)$ at some selected frequency, (*usually* $\omega = 0$; that is, we match *steady-state* gains by choosing $G(0) = D(1)$).

For example, if $G(s) = \dfrac{2}{(s+1)(s+2)}$, then

$$D(z) = \frac{(1-\mathrm{e}^{-T})(1-\mathrm{e}^{-2T})(z+1)^2}{4(z-\mathrm{e}^{-T})(z-\mathrm{e}^{-2T})},$$

(*double* zero for s infinite).

Example 8.4

Design a digital filter equivalent to $G(s) = \dfrac{1}{1+s}$ using

(i) trapezoidal approximation,
(ii) pole-zero matching.

Compare the results from (i) and (ii) with the analog filter output, given that the input is a unit step function and that $T = 0.2$.

(i) Trapezoidal approximation $D(z) = \dfrac{1}{1 + \dfrac{2}{T}\left(\dfrac{z-1}{z+1}\right)}$,

that is, $D(z) = \dfrac{(z+1)}{(z+1) + \dfrac{2}{T}(z-1)}$.

When $\quad T = 0.2,\ D(z) = \dfrac{z+1}{11z-9}$.

Therefore $\dfrac{Z\{c_n\}}{Z\{x_n\}} = \dfrac{z+1}{11z-9}$ where $\{c_n\}$ is the digital computer output.

The corresponding difference equation is

$$c_n = \frac{1}{11}(x_n + x_{n-1} + 9c_{n-1}).$$

For the unit step input, $x_n = 1,\ n \geqslant 0$.
Therefore $\quad c_0 = 0.0909,\quad c_1 = 0.2562,\quad c_2 = 0.3914,$
$\qquad\qquad c_3 = 0.5021,\quad c_4 = 0.5926,\quad c_\infty = 1.$

(ii) Pole-zero matching $D(z) = \left(\dfrac{1-e^{-T}}{2}\right)\left(\dfrac{z+1}{z-e^{-T}}\right)$.

When $T = 0.2,\ D(z) = 0.09063\left(\dfrac{z+1}{z-0.8187}\right)$.

The corresponding difference equation is

$$c_n = 0.09063\,(x_n + x_{n-1}) + 0.8187\,c_{n-1}.$$

For the unit step input, $x_n = 1,\ n \geqslant 0$.
Therefore $\quad c_0 = 0.0906,\quad c_1 = 0.2555,\quad c_2 = 0.3904,$
$\qquad\qquad c_3 = 0.5009,\quad c_4 = 0.5914,\quad c_\infty = 1.$
Comparing the values in (i) and (ii) with the analog filter output $y(t) = 1 - e^{-t}$,
$y(0) = 0,\qquad\qquad y(0.2) = 0.1813,\quad y(0.4) = 0.3297$
$y(0.6) = 0.4512,\qquad y(0.8) = 0.5507,\quad y(\infty) = 1.$

Example 8.5

Design a digital filter equivalent to $G(s) = \dfrac{1}{s^2+1}$ using

(i) the trapezoidal approximation,
(ii) pole-zero matching.

Sketch a graph which compares the outputs of (i) and (ii) with the analog filter

output when the input is a unit step function and $T = \dfrac{\pi}{3}$.

____

The analog filter output is given by

$$y(t) = L^{-1}\left\{\frac{1}{s(s^2+1)}\right\} = L^{-1}\left\{\frac{1}{s} - \frac{s}{s^2+1}\right\}$$

$$= 1 - \cos t.$$

That is, $y(nT) = 1 - \cos n\frac{\pi}{3}$ when $T = \frac{\pi}{3}$.

One cycle of this output will be plotted on the graph and compared with the outputs of the two digital filters.

(i) Trapezoidal approximation $D(z) = \dfrac{1}{\left(\dfrac{2}{T}\left(\dfrac{z-1}{z+1}\right)\right)^2 + 1}$.

That is, $D(z) = \dfrac{\dfrac{T^2}{4}(z+1)^2}{(z-1)^2 + \dfrac{T^2}{4}(z+1)^2}$.

When $T = \dfrac{\pi}{3}$, $D(z) = \dfrac{0.27416(z^2 + 2z + 1)}{1.27416z^2 - 1.45169z + 1.27416}$.

The corresponding difference equation is

$$c_n = 0.21517(x_n + 2x_{n-1} + x_{n-2}) + 1.13933c_{n-1} - c_{n-2}.$$

Therefore $c_0 = 0.2152,$ $c_1 = 0.8907,$ $c_2 = 1.6603,$

$c_3 = 1.8616,$ $c_4 = 1.3214,$ $c_5 = 0.5046,$

$c_6 = 0.1142,$ $c_7 = 0.4862,$ $c_8 = 1.3006.$

(ii) Pole-zero matching $D(z) = \dfrac{(1 - e^{jT})(1 - e^{-jT})(z+1)^2}{4(z - e^{jT})(z - e^{-jT})}$.

That is, $D(z) = \dfrac{(2 - 2\cos T)(z+1)^2}{4(z^2 - 2\cos T \cdot z + 1)}$.

when $T = \dfrac{\pi}{3}$, $D(z) = \dfrac{(z+1)^2}{4(z^2 - z + 1)}$.

The corresponding difference equation is

$$c_n = \frac{1}{4}(x_n + 2x_{n-1} + x_{n-2}) + c_{n-1} - c_{n-2}.$$

Therefore $c_0 = 0.25,$ $c_1 = 1,$ $c_2 = 1.75,$

$c_3 = 1.75,$ $c_4 = 1,$ $c_5 = 0.25,$

$c_6 = 0.25,$ $c_7 = 1,$ $c_8 = 1.75,$

and the sequence repeats itself.

Comparing with the analog output $y(t) = 1 - \cos t$,

$$y(0) = 0, \qquad y\left(\frac{\pi}{3}\right) = 0.5, \qquad y\left(\frac{2\pi}{3}\right) = 1.5,$$

$$y(\pi) = 2, \qquad y\left(\frac{4\pi}{3}\right) = 1.5, \qquad y\left(\frac{5\pi}{3}\right) = 0.5,$$

$$y(2\pi) = 0, \qquad y\left(\frac{7\pi}{3}\right) = 0.5, \qquad y\left(\frac{8\pi}{3}\right) = 1.5, \text{ etc.}$$

The outputs are compared in Fig. 8.6.

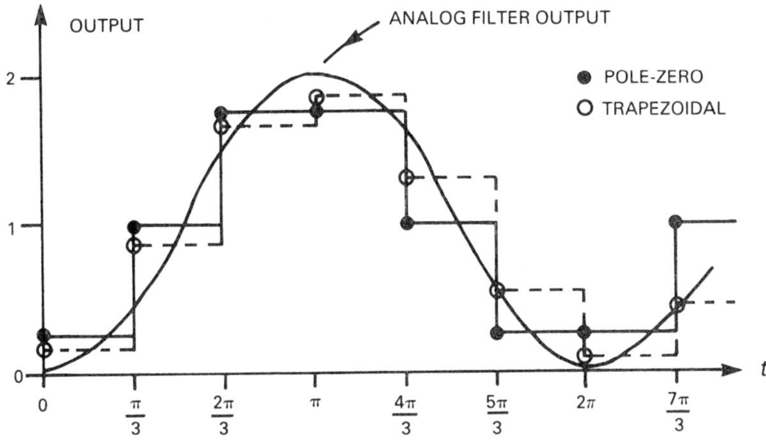

Fig. 8.6

The 'staircase' functions shown in Fig. 8.6 represent the outputs from a zero-order hold reconstruction device, ——— for pole-zero, – – – – for trapezoidal. Note that the pole-zero sequence *leads* the analog output, but that this is compensated by the *lag* due to zero-order hold reconstruction. The trapezoidal results are less accurate, but $T = \dfrac{\pi}{3}$ is rather a large stepsize for this problem.

Problems

(1) Obtain the general formula for c_n in Example 8.4 for both trapezoidal integration and pole-zero matching.

(2) Use the pole-zero matching method to write down the digital filters equivalent to

(i) $\dfrac{10}{s+10}$ (ii) $\dfrac{s+2}{(s+1)(s+3)}$ (iii) $\dfrac{10}{s^2+2s+5}$.

(3) Use the trapezoidal integration method to obtain digital filters equivalent to the analog filters in Problem 2.

(4) (i) Use the pole-zero matching method to obtain a digital filter equivalent to $G(s) = \dfrac{1}{1+s}$ when the sampling interval is $T = \dfrac{\pi}{2}$. If the input function is $\sin t$, find the general formula for the digital computer output sequence $\{c_n\}$. Deduce that as $n \to \infty$ $|c_n| < 0.4586$, to 4 decimal place accuracy.

(ii) Use the trapezoidal approximation method to obtain a digital filter as in (i). Write down the corresponding difference equation relating digital computer output and input. For the input function, $\sin t$, calculate c_n, $n = 0, 1, 2, 3, 4, 5$. Compare with the results in (i) and with the analog filter output.

(5) Simpson's Rule integration procedure is as follows:

$$\int_{(n-2)T}^{nT} f(t)\,dt = \frac{T}{3}\,(f_{n-2} + 4f_{n-1} + f_n).$$

Apply this to the analog filter $\dfrac{a}{s+a}$ with input $x(t)$ and output $y(t)$ to show that the equivalent digital filter is

$$D(z) = \frac{a}{a + \dfrac{3}{T}\left(\dfrac{z^2-1}{z^2+4z+1}\right)}.$$

Demonstrate that the transformation $s = \dfrac{3}{T}\left(\dfrac{z^2-1}{z^2+4z+1}\right)$ does *not* map the stable region of the s-plane into the stable region of the z-plane.

Answers to problems

(1) (i) $c_n = 1 - \dfrac{10}{11}\left(\dfrac{9}{11}\right)^n$, (ii) $c_n = 1 - 0.90937(0.81873)^n$.

(2) (i) $\dfrac{(1-e^{-10T})(z+1)}{2(z-e^{-10T})}$, (ii) $\dfrac{(1-e^{-T})(1-e^{-3T})(z-e^{-2T})(z+1)}{3(1-e^{-2T})(z-e^{-T})(z-e^{-3T})}$,

(iii) $\dfrac{(1-2e^{-T}\cos 2T + e^{-2T})(z+1)^2}{2(z^2 - 2ze^{-T}\cos 2T + e^{-2T})}$.

(3) (i) $\dfrac{10T(z+1)}{10T(z+1)+2(z-1)}$, (ii) $\dfrac{2T^2(z+1)^2+2T(z^2-1)}{3T^2(z+1)^2+8T(z^2-1)+4(z-1)^2}$,

(iii) $\dfrac{10T^2(z+1)^2}{5T^2(z+1)^2+4T(z^2-1)+4(z-1)^2}$.

(4) (i) $\dfrac{0.39606(z+1)}{z-0.20788}$,

$$c_n = 0.3007\sin\frac{n\pi}{2} - 0.4586\cos\frac{n\pi}{2} + 0.4586(0.20788)^n,$$

(ii) $\dfrac{z+1}{(z+1)+1.27324(z-1)}$, $c_n = 0.4399(x_n+x_{n-1})+0.1202\,c_{n-1}$

$c_0 = 0,\qquad c_1 = 0.4399,\qquad c_2 = 0.4928,$
$c_3 = -0.3807,\quad c_4 = -0.4857,\quad c_5 = 0.3815.$
Compare with pole-zero results and analog output,
$c_0 = 0,\qquad c_1 = 0.3961,\qquad c_2 = 0.4784,$
$c_3 = -0.2966,\quad c_4 = -0.4577,\quad c_5 = 0.3009.$

$$y(0) = 0,\qquad y\left(\frac{\pi}{2}\right) = 0.6039\qquad y(\pi) = 0.5216$$

$$y\left(\frac{3\pi}{2}\right) = -0.4955,\quad y(2\pi) = -0.4991,\quad y\left(\frac{5\pi}{2}\right) = 0.5002.$$

REFERENCES

Apostol, T. M., (1957) *Mathematical analysis, a modern approach to advanced calculus,* Addison Wesley

Braun, M., (1975) *Differential equations and their applications.* Springer-Verlag

Burghes, D. N., & Graham, A., (1980) *Introduction to control theory including optimal control.* Ellis Horwood

Gabel, R. A., & Roberts, R. A., (1973) *Signals and linear systems,* John Wiley

Holbrook, J. G., (1966) *Laplace transforms for electronic engineers,* Pergamon Press

Huntley, I. D., & Johnson, R. M., (1983) *Linear and nonlinear differential equations,* Ellis Horwood

Kreyszig, E., (1979) *Advanced engineering mathematics,* John Wiley

Oppenheim, A. V., & Willsky, A. S., (1983) *Signals and systems,* Prentice-Hall

Stanley, W. D., (1975) *Digital signal processing,* Reston

Stearns, S. D., (1975) *Digital signal analysis,* Hayden

Index